D1065314

TECHNIQUES *in* PROTEIN BIOSYNTHESIS

TECHNIQUES *in* PROTEIN BIOSYNTHESIS

Edited by

P. N. CAMPBELL

Department of Biochemistry
University of Leeds
Yorkshire, England

and

J. R. SARGENT

Department of Biological Chemistry
Marischal College
University of Aberdeen
Scotland

VOLUME 2

ACADEMIC PRESS
LONDON · NEW YORK

ACADEMIC PRESS INC. (LONDON) LTD.
Berkeley Square House
Berkeley Square
London, W1X 6BA

U.S. Edition published by
ACADEMIC PRESS INC.
111 Fifth Avenue
New York, New York 10003

Library of Congress Catalog Card Number: 66-30149
SBN: 12-158162-4

Printed in Great Britain by
John Wright & Sons Ltd., at the Stonebridge Press, Bath Road, Bristol

List of Contributors to Volume 2

J. E. ALLENDE, *Departamento de Quimica, Facultad de Ciencias, Universidad de Chile, Santiago, Chile.*

H. BLOEMENDAL, *Department of Biochemistry, University of Nijmegen, The Netherlands.*

W. J. BRAMMAR, *Department of Molecular Biology, University of Edinburgh, Edinburgh, Scotland.*

P. N. CAMPBELL, *Department of Biochemistry, University of Leeds, Yorkshire, England.*

H. NOLL, *Department of Biological Sciences, Northwestern University, Evanston, Illinois, U.S.A.*

P. S. TODD, *Department of Biochemistry, University of Leeds, Yorkshire, England.*

C. VENNEGOOR, *Department of Biochemistry, University of Nijmegen, The Netherlands.*

H. B. WAYNFORTH, *Courtauld Institute of Biochemistry, Middlesex Hospital Medical School, London, England.*

Preface to Volume I

The Walrus and the Carpenter
Were walking close at hand;
They wept like anything to see
Such quantities of sand:
'If only this were cleared away,'
They said, 'it would be grand!'

THE WALRUS AND THE CARPENTER
THROUGH THE LOOKING GLASS
Lewis-Carroll

We suspect that with the present rate of increase in the literature, many research workers may have echoed the sentiments expressed above in the task of keeping abreast of publications. Before undertaking this project, we were conscious of the very rapid rate of progress in protein biosynthesis and the difficulties of writing a book of more than a fleeting impact in this field. What then was our excuse for succumbing to the invitation of the publishers and adding to the already massive literature on the subject?

While the interpretations of the results of experiments do indeed change rapidly, it is probably true to state that the general methods used to obtain the results are more long lasting. Moreover, it is not always easy to obtain from original papers the best method to tackle a particular problem. We thought that a useful contribution could be made by confining a book in the main to techniques. We, therefore, invited as authors scientists who are active workers in the field and we hope that their contributions will help those who are new entrants. We have in mind especially those readers who are not able to learn the techniques at first hand and who may welcome some general guidance to the literature. In this first volume the only chapter which does not deal with techniques is the first which we, as Editors, have written as a general introduction.

We are most grateful to the authors for their cooperation, the various publishers and authors who have given their permission to reproduce original figures, and Academic Press for their cooperation.

P. N. CAMPBELL
J. R. SARGENT

vii

Contents

CHAPTER 1

The Genetic Approach to the Study of Protein Biosynthesis

CHAPTER 2

Protein Biosynthesis in Plant Systems

CHAPTER 3

Polysomes: Analysis of Structure and Function

CHAPTER 4

Fractionation of Ribosomal Proteins

CHAPTER 5

Animal Operative Techniques
(In the Mouse, Rat, Guinea Pig and Rabbit)

APPENDIX

The Use of "High Energy" Phosphate Compounds in "in Vitro" Studies on Protein Synthesis

CHAPTER 1

The Genetic Approach to the Study of Protein Biosynthesis

W. J. BRAMMAR

*Department of Molecular Biology, University of Edinburgh,
Edinburgh, Scotland*

I. INTRODUCTION

It is the aim of this review to consider the contribution of genetic techniques to our understanding of the overall process of protein biosynthesis. Our current knowledge of this process has been largely derived from biochemical investigations, and the contribution of genetics has been relatively meagre. That this is so is due to the lack of a simple methodology for the isolation of mutants in which the central processes of protein synthesis are impaired. Recently, however, several different approaches have been developed and mutants with altered ribosomes, transfer RNAs and aminoacyl-tRNA synthetases have been isolated and studied in detail. The combination of biochemical and genetical approaches will obviously be enormously productive in the study of protein synthesis in the near future. The great contribution of genetics to the elucidation of the genetic code itself, and to the understanding of the regulation of protein biosynthesis, will be included in the review.

II. THE GENETIC CODE

A. General Nature of the Genetic Code

The amino acid sequence of a protein is specified by the nucleotide sequence of the DNA of the corresponding structural gene. Since there are twenty different amino acids in proteins, but only four different nucleotides in DNA, some combination of nucleotides, larger than a doublet, must correspond to each amino acid. The elucidation of the structure of DNA in 1953 (Watson and Crick, 1953) stimulated much interest in the coding problem and much ingenious speculation on the structure of the code resulted. The early work on the coding problem has been extensively reviewed (e.g. Crick, 1963, 1966; Woese, 1967).

The general nature of the genetic code was finally placed on a firm empirical foundation by the brilliant exploitation of acridine mutants of the *rII* system of coli-phage T_4 (Crick *et al.*, 1961). Acridine-induced mutants, which are generally non-leaky (i.e. they result in a completely inactive gene-product), cannot be induced to revert by base-analogue mutagens. When mutants of this type *do* revert, they often do so by a second mutation at a different site. It was postulated, therefore, that the mutagenic effect of acridines was due to the insertion or deletion of base-pairs in DNA, rather than to alteration of a base-pair (Brenner *et al.*, 1961). Study of acridine mutants in the *rIIB* cistron of the phage showed that all these mutants could be divided into two classes, arbitrarily called (+) and (−) types. When doubly altered strains, containing one (+) and one (−) mutation, were prepared, they frequently had a non-mutant ("pseudowild") phenotype. Reconstituted stocks with three (+) or three (−) mutations were pseudowild, while those with two mutations of the same sign had a mutant phenotype. In explanation of these findings it was proposed that the nucleotide sequence of the DNA is read sequentially, three nucleotides at a time, from a fixed starting point. Acridine mutation, adding or deleting a nucleotide, produces a change in the phase of reading beyond the site of the mutational event. The phase can be corrected, and gene-function restored, by a second mutation, opposite in sense to the first. The nucleotide sequence between the mutational sites would still be out of phase and the corresponding amino acids would differ from the wild-type sequence in that region. When three nucleotides were added or deleted, the correct reading phase would be restored beyond the site of the last change and the protein would contain one amino acid more or less than the wild-type protein (Crick et al., 1961). These predictions have now been directly confirmed by studies of amino acid changes produced by phase-shift mutations in the lysozyme gene of phage T_4 (Streisinger et al., 1966). In some cases (+ −) pairs were still inactive, due to the presence of "barriers" in the shifted reading between the two mutations. These barriers can be reverted by base analogue mutagens, showing that they are due to localized unacceptable points in the base sequence. The original observations have now been extended to include quadruple and quintuple (+) mutants, which are inactive, and a sextuple (+) mutant which again has a pseudowild phenotype (Brenner, 1966). [For detailed methodology of phase-shift mutation in the T_4 rIIB system see Barnett et al. (1967).]

It is evident from these results that quite long sequences of random triplets generated by phase-shifts contain relatively few unacceptable codons, and, therefore, that the code must be degenerate. The question of whether the code is overlapping was not solved by these experiments. However, overlapping codes are virtually eliminated by the evidence from studies with many different systems that mutations lead only to single amino acid changes. Thus the evidence points towards a non-overlapping, degenerate, triplet code, which is read sequentially from a fixed starting point.

B. Colinearity

I. Colinearity of Genetic Map and Protein

The arrangement of nucleotides in a gene determines the sequence of a linear array of amino acids in a polypeptide chain. It had long been suspected that the gene and the protein which it codes were colinear.

2

This expectation was convincingly demonstrated by two groups of workers in 1964. Yanofsky *et al.* (1964) worked with missense mutants in the *A*-gene of the tryptophan operon of *Escherichia coli*. This system is particularly suitable for a study of amino acid replacements because of the existence of a simple assay for mutant A-proteins, based on their stimulation of B-protein activity (Smith and Yanofsky, 1963). Thus mutant A-proteins, themselves inactive in tryptophan synthesis, are nevertheless active in stimulating the conversion of indole and serine to tryptophan catalysed by the B-protein. The mutants were mapped by transduction with the generalized transducing phage Plkc, using a histidine marker as an internal standard of transduction frequency. The amino acid alterations produced by each mutant were determined by chemical analysis of the appropriate region of the protein.

The altered regions were placed in order by determining the amino acid sequence of a large part of the protein. The order of the amino acid substitutions in the polypeptide chain was the same as the order of the mutants on the genetic map. In addition, distances measured by recombination were proportional to the physical distance on the polypeptide chain (Yanofsky *et al.*, 1964). The amino acid sequence of the A-protein has subsequently been completed and the demonstration of colinearity extended to cover 186 of the 267 residues (Yanofsky *et al.*, 1967). The data showing colinearity of the *E. coli trp A* gene with the tryptophan synthetase A-protein are reproduced in Fig. 1.

Colinearity was also demonstrated in a different way using mutants in the head protein gene of bacteriophage T_4 (Sarabhai *et al.*, 1964). This study exploited the properties of a set of "conditional lethal mutants", the amber mutants, distinguished by their response to a specific set of suppressors. These mutants grow normally on a strain of *E. coli* carrying the relevant suppressor gene (su^+) and fail to grow on a suppressor-free (su^-) strain. Amber mutants of the head protein of phage T_4 produce fragments of the polypeptide chain when grown in the su^- strain (Sarabhai *et al.*, 1964). About 90% of the protein of the mature phage particle is head protein (Van Vunakis *et al.*, 1958; Brenner *et al.*, 1959) and 60–70% of the proteins synthesized late during infection are incorporated into mature phage particles (Koch and Hershey, 1959). Thus, since more than half of the late protein synthesis of the infected cell is synthesis of head protein, fragments of the head protein chain can be detected without prior purification.

Several mutants, induced by base-analogue mutagens, were located in the head protein cistron. Ten different sites were found and one mutant from each was studied. Double mutants were constructed by

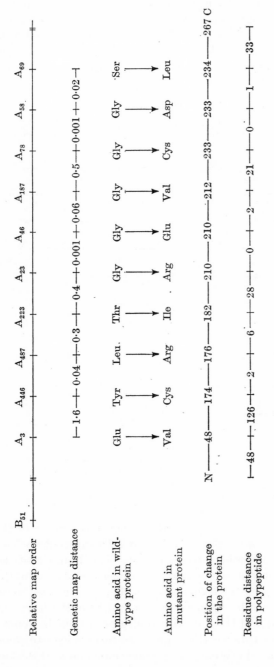

FIG. 1. Colinearity of the *trp* A-gene and the tryptophan synthetase A-protein of *E. coli* (taken from Yanofsky *et al.*, 1967). The genetic map was prepared by two- and three-factor crosses using transduction with the phage P1. The genetic map distance is the percentage recombination frequency between a pair of markers. Amino acid replacements were determined by isolating and sequencing the relevant tryptic or chymotryptic peptides. There is a good correlation between genetic map distance and residue distance in the polypeptide chain, with an average value of 0·015 map units per amino acid residue.

pair-wise crosses, so that both two-factor and three-factor crosses could be used to order the mutants.

The proteins being synthesized in the phage-infected cell were labelled, the cells lysed, nucleic acid removed and the mixture digested with trypsin or chymotrypsin. The resulting peptides were characterized

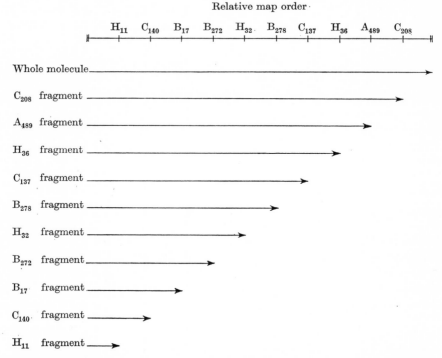

FIG. 2. Colinearity between gene and protein in the head-protein system of phage T$_4$ (based on Sarabhai *et al.*, 1964). The hierarchical ordering of the prematurely terminated polypeptide fragments produced by this set of amber mutants corresponds with the order of the mutations on the genetic map.

by high-voltage ionophoresis and radioautography. Some of the peptides of the head protein were present in cells infected with a given amber mutant, while others were absent. The number of peptides found in any one mutant can be taken as a measure of the length of the fragment produced and the fragments can then be arranged in order of increasing size. The order deduced in this way from the fragments was the same as the order of the mutational alterations on the genetic map (Fig. 2). Thus it was shown that amber mutations lead to premature termination of the head protein polypeptide chain and that the locations of the termini produced in different mutants are colinear with the genetic map (Sarabhai *et al.*, 1964).

Evidence obtained with phase-shift mutants has recently demonstrated the colinearity of the genetic map of the T_4 e-gene and the polypeptide chain of T_4 lysozyme (Streisinger *et al.*, 1966).

2. Colinearity of Genetic Map and DNA Structure

Several recent studies have established the colinearity of genetic maps and DNA structure. Hogness *et al.* (1966) have determined the order of genetic markers in a phage λ-DNA molecule by assaying the markers associated with separated DNA fragments of different known lengths generated by hydrodynamic shear. The assay procedure involves infection of an *E. coli* strain, using a multiply marked helper phage (Kaiser and Hogness, 1960). The order of six genes determined in this way was: (N, i^{λ})-O-P-Q-R-*end*, where *"end"* represents one of the cohesive ends of the vegetative λ-DNA molecule which is necessary for activity in this assay system. This is precisely the order of the markers on the genetic map of the vegetative phage. The order of the three genes of the galactose operon located on the left half of the DNA of λ*dg* (i.e. the defective λ phage which transduces the *gal* genes) was shown to be: (*Left end*)-*k*-*t*-*e*, the order expected on the basis of Campbell's model for the formation of λ*dg* (Campbell, 1962).

Mosig (1966) determined the physical location of genetic markers on the T_4 DNA molecule using a class of light-density phage particles containing DNA fragments two-thirds the length of normal T_4 DNA. The frequency with which light particles contributed two markers in mixed infection with normal particles was used as a measure of distance between the markers. The physical map of the markers obtained in this way agreed very well with the genetic map of T_4. Goldberg (1966) has employed "marker rescue" experiments with T_4 DNA fragments to estimate the physical distances separating markers on the phage DNA. Fragmented DNA was added to *E. coli* spheroplasts which were subsequently infected with intact mutant phage. In this way the joint transformation frequency of a pair of markers could be studied as a function of the extent of fragmentation of the DNA molecule. The results obtained with this method are consistent with colinearity of T_4 DNA and its genetic map.

C. The Genetic Code Today

The first codon assignment was determined in 1961, when Nirenberg and Matthaei (1961) observed that polyU specifically stimulated the synthesis of polyphenylalanine in an *in vitro* protein synthesizing system derived from *E. coli*. Thanks largely to the development of the triplet binding test (Nirenberg and Leder, 1964), and to techniques for producing polynucleotides with accurate repeating sequences (Nishimura *et al.*, 1965a, b), the codons for all twenty amino acids are now

*UpUpU / *UpUpC PHE	*UpCpU / *UpCpC SER	*UpApU / *UpApC TYR	*UpGpU / *UpGpC CYS
UpUpA / *UpUpG LEU	*UpCpA / *UpCpG SER	UpApA OCHRE / UpApG AMBER	UpGpA C.T. / *UpGpG TRP
CpUpU / CpUpC LEU	*CpCpU / *CpCpC PRO	*CpApU / *CpApC HIS	*CpGpU / *CpGpC ARG
CpUpA / *CpUpG LEU	*CpCpA / *CpCpG PRO	*CpApA / *CpApG GLN	*CpGpA / *CpGpG ARG
*ApUpU / *ApUpC ILE	*ApCpU / *ApCpC THR	*ApApU / *ApApC ASN	ApGpU / *ApGpC SER
ApUpA ILE / *ApUpG MET	*ApCpA / *ApCpG THR	*ApApA / *ApApG LYS	ApGpA / *ApGpG ARG
*GpUpU / *GpUpC VAL	*GpCpU / *GpCpC ALA	*GpApU / *GpApC ASP	*GpGpU / *GpGpC GLY
*GpUpA / *GpUpG VAL	*GpCpA / *GpCpG ALA	*GpApA / *GpApG GLU	*GpGpA / *GpGpG GLY

Fɪɢ. 3. The genetic code. Triplets marked with an asterisk give a positive triplet binding test (Nirenberg and Leder, 1964): those underlined have been determined by the use of synthetic polynucleotides as messengers. The codons marked ochre, amber and C.T. act as signals for the termination of polypeptide chain synthesis. Most of the evidence on which these allocations are based comes from *E. coli*.

known with a high degree of certainty. The genetic code as it is conventionally displayed is shown in Fig. 3. The codons determined by the triplet binding test are marked with an asterisk: those assigned by the use of synthetic mRNAs are underlined. The use of these chemical techniques for codon determination has been covered in Vol. 1 of this series by Bretscher and Jones (1967). The binding test cannot always be relied upon to give an unambiguous assignment. Some triplets fail to give a positive binding, while others stimulate the binding of tRNAs for more than one amino acid. The use of synthetic polynucleotides of repeating sequence to direct the synthesis of polypeptides appears to be a very reliable, although technically difficult, method for codon determination. It has been particularly useful for assigning those codons which fail to give a positive binding test.

D. Approaches to Codons used in vivo

Both the binding method and the synthetic messenger method for codon determination are open to the objection that they are used in a cell-free system and, therefore, may not be free from artefacts. The assignments reached by *in vitro* experiments can be tested by several techniques using intact cells. The degree of agreement between the results obtained *in vitro* and *in vivo* is so high that we may be fairly confident that the current codon assignments are correct.

I. Amino Acid Replacements in Mutants

Much of the *in vivo* information on codon assignments has come from analyses of amino acid replacements occurring in proteins as a result of mutations. Three proteins, human haemoglobin, tobacco mosaic virus protein and the A-protein of tryptophan synthetase of *E. coli*, have been particularly useful in studies of this nature.

Most abnormal human haemoglobins have been recognized because they are electrophoretically different from normal adult haemoglobin. The amino acid changes observed with this system, therefore, are likely to involve a charge change. The data for 36 different abnormal human haemoglobins have been summarized by Lehmann and Huntsman (1966). All 36 replacements are consistent with a single base change according to the current code.

Tobacco mosaic virus has single-stranded RNA as its genetic material. Most of the mutants studied with this system were produced by treating the whole virus or its RNA with nitrous acid. After reconstitution with untreated coat protein, the virus is inoculated onto a host where local lesions are produced, single lesions isolated and these are transferred to another host to give systemic infections. Mutants can be distinguished by the appearance of unusual numbers or shapes of lesions on the local-lesion host, or by unusually mild or severe symptoms on the systemic hosts.

Nitrous acid converts cytosine to uracil, and adenine to hypoxanthine, which is then copied as if it were guanine. The great majority of amino acid replacements observed could fit with those expected from the mutagen specificity. The few that do not fit are probably spontaneous mutations picked up with the nitrous acid mutants (review, Wittmann and Wittmann-Liebold, 1966).

Yanofsky and his colleagues have studied amino acid replacements in the tryptophan synthetase A-protein of *E. coli* (Yanofsky *et al.*, 1966b). As mentioned previously, this system has the fortunate advantage that mutant A-proteins, although inactive in the normal biosynthesis of tryptophan, are nevertheless active in stimulating the conversion of indole plus serine to tryptophan, catalysed by the B-protein (Smith and Yanofsky, 1963). This reaction can then be used as a convenient assay to monitor the purification of mutant A-proteins. In addition, there are several sensitive techniques for distinguishing different back-mutations (Allen and Yanofsky, 1963), making it possible to study several amino acid replacements at a given site. All but one of the amino acid replacements found are consistent with a single base-change on current codon assignments. One replacement, glutamic acid (GAG) by methionine (AUG), implies a change of at least two bases (Drapeau *et al.*, 1968). This mutation has been picked up only once and has never reverted, which supports the interpretation of a change of two bases.

The agreement between the observed amino acid replacements *in vivo* and the *in vitro* studies using triplet-binding and synthetic messengers is so good that we may be confident of the current codon assignments. Assuming that these assignments are the correct ones, we can use amino acid replacement data to determine some of the codons used *in vivo*. Because of the high degeneracy of the code, most amino acid replacements do not define a unique codon. Methionine and tryptophan have unambiguous codons, however, and the nonsense codons UAA (ochre), UAG (amber) and UGA are also unambiguous. We can thus define any codon which is related to a methionine, tryptophan or nonsense codon, unless it is related by a base-change in the degenerate "third position".

There are six amino acids which have two assigned codons related to the amber (UAG) and ochre (UAA) codons: lysine (AAG and AAA), glutamine (CAG and CAA), glutamic acid (GAG and GAA), serine (UCG and UCA), leucine (UUG and UUA) and tyrosine (UAU and UAC). Weigert *et al.* (1966), using the alkaline phosphatase system of *E. coli*, have shown that both codons for the first five of these amino acids can function *in vivo*, since each amino acid occurs as a substitution resulting from mutational alteration of both UAG and UAA nonsense codons. (This conclusion cannot be drawn for the tyrosine codons, which are related to UAG and UAA by a change in the degenerate third position.) There appear to be differences in the frequencies with which alternative codons are used, however. Out of 25 amber mutants screened, none arose by mutation of a lysine codon: of 24 ochre mutants, seven arose from a lysine codon. Thus, although both AAA and AAG can act as lysine codons in *E. coli*, AAA appears to be much more frequently used (Weigert *et al.*, 1966).

Much information on codons utilizable *in vivo* has been derived from an extensive study of amino acid replacements at one particular site in the tryptophan synthetase A-protein of *E. coli* (Yanofsky *et al.*, 1966b). Two independent mutations at the same site result in the replacement of the wild-type glycine residue at position 210 by arginine (mutant *trp A23*) and glutamic acid (*trp A46*). *A23* yields revertants in which the mutant arginine residue is replaced by glycine, serine, threonine and isoleucine residues. This latter replacement, arginine by isoleucine, determines the corresponding codons as AGA (arg) and AUA (ile) and thus the wild-type glycine codon as GGA. Mutant *A46* yields revertants in which the glutamic acid is replaced by glycine (GGA), alanine (GCA) and valine (GUA) residues.

Berger and Yanofsky (1967) have extended this analysis by using a missense suppressor to select for the replacement of the glutamic acid of mutant *A46* by aspartic acid. This change would not normally be observable, since aspartic acid at residue 210 gives a mutant phenotype. In the presence of a specific missense suppressor gene, however, the aspartic acid codon is misread as a glycine codon, resulting in a "pseudowild" phenotype. After removal of the mutant *A*-gene to a non-suppressing background, the A-protein from the new mutant (*A46*[asp]) was isolated, and shown to contain aspartic acid as residue 210. *A46*[asp] was further reverted to wild-type and "pseudowild" types and revertants containing glycine, alanine, valine and asparagine at residue 210 were obtained (Berger *et al.*, 1968).

The mutant aspartic acid codon of *A46*[asp], being derived from the glutamic acid codon GAA by a change in the third position, could be either GAU or GAC. However, analysis of the altered sequences of the A-proteins produced by strains containing complementary phase-shift mutations covering this part of the *A*-gene defines the mutant aspartic acid codon as GAU (Berger *et al.*, 1968) (see Fig. 6). Further, Treffers' mutator gene, which stimulates only A/T → C/G transversions (Yanofsky *et al.*, 1966a), has been used to produce the mutation corresponding to the replacement of glutamic acid by aspartic acid. The aspartic acid codon produced in this way must be GAC. This codon also gives rise to glycine (GGC), alanine (GCC), valine (GUC) and asparagine (AAC) codons on reversion (Berger *et al.*, 1968). The amino acid and codon replacements obtained at residue 210 of the tryptophan synthetase of *E. coli* are summarized in Fig. 4.

2. Amino Acid Changes Produced by Phase-shift Mutations

The genetic analysis of acridine-induced mutants in the *rIIB* cistron of phage T_4 led to the conclusion that the message RNA is translated sequentially, three bases at a time, and from a fixed starting

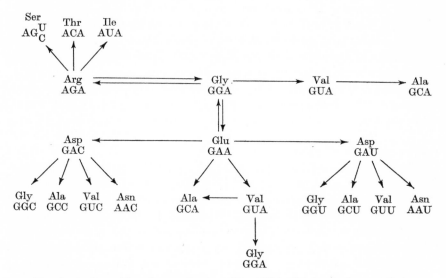

Fig. 4. Amino acid replacements at position 210 in the tryptophan synthetase A-protein (*E. coli*) and probable corresponding codon assignments. A total of 19 different amino acid replacements have been observed, including 10 amino acids and 17 different codons (taken from Berger *et al.*, 1968).

point (Crick *et al.*, 1961). This being so, the addition of a nucleotide at any point will throw the subsequent reading out of phase. The subtraction of a nucleotide at a different point will restore the original phase of reading, except in the region between the two alterations (Crick *et al.*, 1961).

This conclusion has now been directly confirmed by Streisinger and colleagues in studies using the lysozyme of phage T_4 (Terzaghi *et al.*, 1966; Okada *et al.*, 1966; Streisinger *et al.*, 1966; Tsugita *et al.*, 1967; Inouye *et al.*, 1967). Independent proflavine-induced mutations of the lysozyme gene were isolated and crossed together in pairs. Certain pairs of mutants yielded both wild and pseudowild progeny. The lysozymes from bacteria infected with wild-type and with a pseudo-wild strain were purified and analysed by the techniques of protein chemistry. In several cases the primary structures of the lysozymes of the wild-type phage and the double-mutant phage differed by a sequence of several contiguous amino acids. Using current codon assignments, it was possible in each case to assign a sequence of bases that could code for the wild-type sequence of amino acids; with the appropriate addition and/or deletion of bases, the assigned base sequence could also code for the changed sequence of amino acids in the mutant strains. An example of the changes observed is shown in Fig. 5.

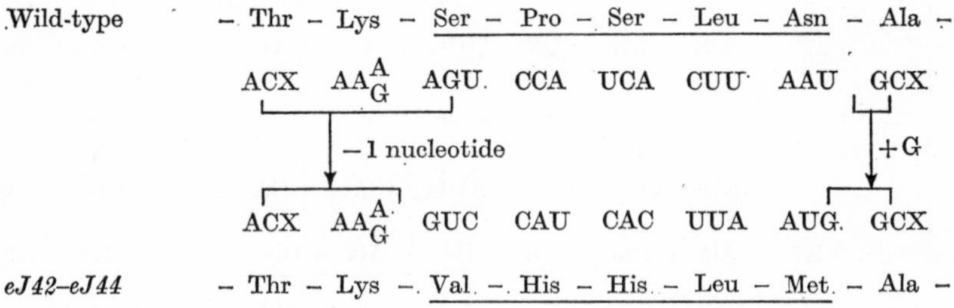

FIG. 5. Changed sequences of amino acids in the lysozyme produced by strains of phage T_4 carrying two phase-shift mutations in the lysozyme structural gene. The codons are those which are compatible with the amino acid sequences determined experimentally (X represents U, C, A or G) (from Streisinger et al., 1966).

It was shown that nucleotide triplets compatible with the two sequences can be derived only if the direction of reading of the triplets (and therefore the mRNA) is from 5′ to 3′, where the 5′ end corresponds to the N-terminal end of the polypeptide chain (Terzaghi et al., 1966).

Assuming that the codons derived from in vitro studies are used in vivo, these studies have shown that the following codons are utilized in T_4-infected E. coli:

Wild-type: Ser-AGU, UCA; Pro-CCA; Leu-CUU; Asn-AAU; Tyr-UAU; Trp-UGG.

Mutants: Leu-UUA; Val-GUC, GUA; His-CAU, CAC; Lys-AAA, AAG; Cys-UGU; Ile-AUA; Met-AUG; Arg-AGG; Glu-GAA.

Frame-shift mutations have now been analysed which consist of both the addition and deletion of a single base or of two bases (Inouye et al., 1967).

A similar analysis has recently been made of altered amino acid sequences present in pseudowild strains carrying two phase-shift mutations in the tryptophan synthetase A-gene of E. coli (Brammar *et al.*, 1967). In this system the pseudowild strains are produced by spontaneous or mutagen-induced reversion of a frame-shift mutant. [Bacterial phase-shift mutants can be recognized by their lack of response to suppressor mutations, by failure to be reverted by base-substitution mutagens and by a positive response to reversion by ICR-compounds* (Whitfield *et al.*, 1966).]

A combination of missense mutation and phase-shift mutation has been very useful for *in vivo* codon determination at a particular site in tryptophan synthetase *A*-gene of *E. coli* (Berger *et al.*, 1968). Amongst the spontaneous revertants of a tryptophan *A*-gene missense mutant,

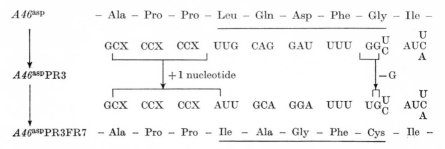

FIG. 6. The altered amino acid sequence in the tryptophan synthetase A-protein (*E. coli*) produced by strain *A46*[asp]PR3FR7, which is derived by two mutational events from the missense mutant strain *A46*[asp]. The codon assignments, which are those compatible with the two amino acid sequences, include GAU for the mutant aspartic acid codon of *A46*[asp] (after Berger *et al.*, 1968).

A46[asp], was a slow-growing strain which proved to contain a phase-shift mutation separated by a few nucleotides from the original missense mutation. The A-protein could not be isolated from this double mutant. A further mutational event, presumably a second phase-shifting event, resulted in a wild phenotype. The A-protein from the phenotypically wild triple mutant differed from wild-type A-protein by a sequence of five contiguous amino acids, as shown in Fig. 6. The codon assignments consistent with the amino acid replacements are shown in Fig. 6 and include the mutant aspartic acid codon (GAU) of the missense mutant *A46*[asp]. This mutant has been shown to give rise to revertants containing glycine (GGU), alanine (GCU), valine (GUU) and glutamine (AAU) at position 210 in the A-protein (Berger *et al.*, 1968). The phase-shift event giving rise to the slow-growing revertant

* Phase-shift mutations are induced in phage by acridines (Crick *et al.*, 1961). Although these compounds are much less effective mutagens for bacteria, certain derivatives of acridine, and especially acridine half mustards, are potent mutagens for bacteria. These derivatives are known as ICR-compounds since they were first synthesized and used at the Institute for Cancer Research, Philadelphia, Pennsylvania.

strain is the base addition, since two independent revertants of this strain have this event in common, but different nucleotide-deletion events (Berger et al., 1968).

The data on amino acid replacements following mutational events are entirely consistent with the codon assignments from in vitro studies. At the present time, 45 codons have been shown to work in vivo and 19

UUU[1]	Phe			UAU[3,10]	Tyr	UGU[3,4]	Cys
						UGC[3]	Cys
UUA[4,5]	Leu	UCA[3,4,5]	Ser				
UUG[1,5]	Leu	UCG[5]	Ser			UGG[5,7,10]	Trp
CUU[3,4]	Leu			CAU[4]	His		
				CAC[3,4]	His		
		CCA[4]	Pro	CAA[5]	Glu	CGA[3]	Arg
CUG[3]	Leu			CAG[1,5,6,7]	Glu	CGG[3]	Arg
AUU[1]	Ile			AAU[1,4]	Asn	AGU[4]	Ser
		ACC[3]	Thr	AAC[1,3]	Asn	AGC[2]	Ser
AUA[2,10]	Ile	ACA[2]	Thr	AAA[4,5]	Lys	AGA[2]	Arg
AUG[6,8,10]	Met			AAG[4,5]	Lys	AGG[9]	Arg
GUU[1]	Val	GCU[1]	Ala	GAU[1]	Asp	GGU[1]	Gly
GUC[1,4]	Val	GCC[1]	Ala	GAC[1]	Asp	GGC[1]	Gly
GUA[2,10]	Val	GCA[1,2]	Ala	GAA[2,5,9]	Glu	GGA[1,2,3]	Gly
GUG[1,6,8]	Val			GAG[3,5,6]	Glu		

1. Berger et al. (1968).
2. Yanofsky et al. (1966b).
3. Brammar et al. (1967).
4. Streisinger et al. (1966).
5. Weigert et al. (1966).

6. Drapeau et al. (1968).
7. Stretton et al. (1966).
8. Berger and Yanofsky (1967).
9. Inouye et al. (1967).
10. Tsugita et al. (1967).

FIG. 7. The amino acid codon assignments which have been shown to be functional in vivo in E. coli or T$_4$-infected E. coli.

to be present in wild-type E. coli or T$_4$-infected E. coli (Fig. 7). Aspects of the code remaining to be elucidated include the genetic signals for starting and stopping transcription and translation and the normal functions, if any, of the three nonsense codons.

E. Directions of Reading

The synthesis of the growing polypeptide chain has been shown to proceed from amino terminus to carboxyl terminus (Bishop et al., 1960). Several lines of evidence now support the idea that the $5' \rightarrow 3'$ direction of the mRNA corresponds to the $N \rightarrow C$ direction of the polypeptide chain.

One line of evidence comes from studies of cell-free protein synthesis using synthetic polynucleotides having a codon of specified sequence at one end of the chain. When ApApA.....pApApC is used as messenger in a cell-free system from either *E. coli* (Salas *et al.*, 1965) or rabbit reticulocytes (Lamfrom *et al.*, 1966), the product is polylysine with C-terminal asparagine and N-terminal lysine. Thus in both these systems translation proceeds from the 5' to the 3' end. Similar observations with an *E. coli* system have been made by Thach *et al.* (1965).

Analyses of amino acid replacements produced by pairs of phase-shift mutations support the conclusion of 5' to 3' translation. By comparing the wild-type amino acid sequence with the altered sequence in the double mutant, it is possible to deduce the probable base sequence for the relevant region of messenger RNA (Streisinger *et al.*, 1966; Brammar *et al.*, 1967). In an exercise of this kind, Terzaghi *et al.* (1966) showed that no solution could be found unless the 5' → 3' direction of the codons parallels the N → C direction of the polypeptide chain.

Guest and Yanofsky (1966) arrived at a similar conclusion by ordering the positions of two different nucleotide changes within the same codon relative to an outside marker. *E. coli trp* A-gene mutants *A23* and *A46* have replacements of the same glycine residue by arginine and glutamic acid, respectively (Helinski and Yanofsky, 1962; Henning and Yanofsky, 1962). When *A23* and *A46* are crossed by transduction, rare intra-codon recombinants can be obtained in which the wild-type amino acid, glycine, has been restored (Guest and Yanofsky, 1965). By incorporating a third closely linked marker, *trp E⁻*, into the cross, and scoring this non-selected marker amongst the recombinants, the *A23* and *A46* mutations can be ordered. Because of the very low frequency of intra-codon recombination, it was necessary to reduce interference from reversion by using a loosely linked marker, *cys⁻*, in the recipient. The arrangement for the ordering of *A23* and *A46* is shown in Fig. 8. The results of such a cross were only consistent with the order *E–A23–A46*. Since *A23* involves a replacement in the 5' end of the mRNA codon, and *A46* affects the middle position, it follows that the order of the codon with respect to *E* must be *E*-5'–3'. It has previously been shown that the *E*-proximal end of the *A*-gene codes for the amino end of the A-protein (Yanofsky *et al.*, 1964), so that the 5' → 3' direction of the codons within the message must correspond to the N → C direction of the A-protein. It is known that mRNA synthesis proceeds from the 5' to 3' end of the polyribonucleotide chain (Goldstein *et al.*, 1965; Maitra and Hurwitz, 1965; Bremer *et al.*, 1965). Since the *trp* operator region is at the *E*-gene end (Matsushiro *et al.*, 1965), it follows that the transcription and translation start from the operator end (Guest and Yanofsky, 1966). There is independent biochemical evidence that the synthesis of the *trp*-operon mRNA starts at the *E*-gene end (Imamoto, *et al.*, 1965).

The *E. coli* β-galactosidase protein has been oriented with respect to the genetic map by immunological studies on *z*-gene nonsense mutants (Fowler and Zabin, 1966). Certain of these mutants produce poly-peptides which cross-react with antiserum to the purified wild-type enzyme. The approximate sizes of the cross-reacting materials, deter-mined by sedimentation in sucrose gradients, fitted well with the map

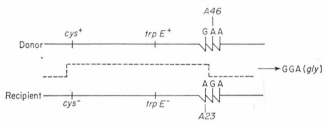

FIG. 8. Arrangement for the ordering of *E. coli trp A*-gene mutants *A23* and *A46* relative to a *trp E* outside marker. The *cys⁻* recipient is used to eliminate the inter-ference due to reversion of the *trp A* marker in the recipient, which would obscure the very low intra-codon recombination (after Guest and Yanofsky, 1966).

distance from the operator end of the *z*-gene to the site of mutation. These studies demonstrate the colinearity of the *z*-gene genetic map and the β-galactosidase polypeptide chain and also orient the chain with respect to the map. The region corresponding to the N-terminal end is closest to the operator region (Fowler and Zabin, 1966). This is sup-ported by the demonstration that the purified polypeptide chain from one particular ochre mutant has the wild-type N-terminal sequence but lacks the normal C-terminus (Brown *et al.*, 1967).

An *E. coli F'lac* episome with a temperature-sensitive replication system has been described (Cuzin and Jacob, 1964). Strains carrying such an episome together with a *lac* deletion on the chromosome become *lac⁻* at high temperature, when the replication of the episome is impaired. The selection of lactose-fermenting revertants at 42° results in strains in which the episome has become integrated into the bacterial chromosome. Beckwith and Signer (1966), using this technique, obtained derivatives in which the newly integrated *lac* operons had both possible orientations relative to the rest of the bacterial chromosome. Since the direction of reading of the sense strand of the DNA duplex must be 3′→5′ (i.e. antiparallel to that the messenger RNA), it follows that the *lac* operons with opposite orientations must be read from opposite DNA strands. The *lac* operon functions equally well on either strand (Beckwith and Signer, 1966).

The direction of reading relative to the circular linkage map has been determined for a number of genes of phage T$_4$. For gene *23*, the structural gene for T$_4$ head protein (Sarabhai *et al.*, 1964), the direction of reading was determined directly, by examining the peptides synthesized by amber mutants of T$_4$ (Sarabhai *et al.*, 1964). The directions

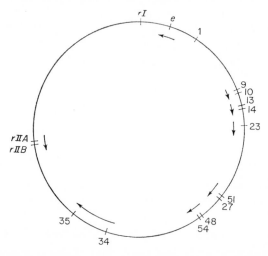

Fig. 9. Certain of the genes of the circular linkage map of phage T$_4$: the arrows indicate the direction of reading. The directions of reading of gene 23 (the structural gene of the T$_4$ head protein) and gene *e* (the structural gene for T$_4$ lysozyme) were determined by analysis of the primary structures of mutant proteins. The directions of reading of the other genes were deduced from the direction of polarity of amber and other mutations (after Streisinger *et al.*, 1968).

of translation of several genes have been deduced from the direction of polarity of amber mutations with respect to neighbouring genes (Stahl *et al.*, 1966). The direction of reading of the *rIIA*- and *B*-genes was determined on the basis of the effect of frame-shift mutations in *A* on the function of *B*, when *B* is joined to *A* by a deletion that removed the intergenic boundary (Crick *et al.*, 1961). The direction of reading of the lysozyme gene (*e*) has been determined by ordering several mutations with respect to outside markers and determining the altered amino acid sequences produced by these mutations within the known sequence of the lysozyme polypeptide chain (Streisinger *et al.*, 1968). The overall results with T$_4$ (Fig. 9) show that lysozyme, which appears relatively late after infection, is read in the same direction as the *rIIB* product, which appears early, and in the opposite direction to other genes whose products appear late. Thus there is no apparent correlation between the time of translation and its direction (Streisinger et al., 1968).

III. NONSENSE CODONS AND CHAIN TERMINATION

A. Nonsense Codons

The in vitro studies of Nirenberg and Khorana and their co-workers have elucidated the structure of the genetic code. It is now known that nearly all of the 64 possible triplets code for an amino acid. Triplets which do not code for an amino acid have been termed "nonsense triplets". The number of nonsense triplets and their structures have been determined by genetic methods.

The amber class of conditional lethal mutants of T_4 form plaques on E. coli CR63, but not on E. coli B (Epstein et al., 1963). Benzer and Champe (1962) showed that amber mutants of the rIIA cistron completely abolished B-function when combined with the deletion r1589, which itself leaves B-function intact.* They suggested that this was because the amber mutant interrupted reading of the message due to presence of a nonsense triplet.

Amber mutations in the head-protein cistron of T_4 produce an amino-terminal fragment of the polypeptide chain (Stretton and Brenner, 1965). Since proteins are synthesized from their amino-termini, the fragment can be produced only by termination of the growing chain at the site of mutation.

A second class of suppressible mutants, the ochre mutants, has been defined. Like amber mutants, ochres also lead to interruption of the reading of the genetic message (Brenner and Beckwith, 1965). Ochre mutants are not suppressed by amber suppressors but ochre suppressors do suppress amber mutants. It has not been formally proved that ochre mutants result in chain termination but the similarity in behaviour of ochre and amber mutants makes this very likely.

Several lines of evidence suggest that nonsense mutations act on the translation of the messenger RNA, rather than on its synthesis or premature breakdown. Nonsense mutants can be phenotypically suppressed by 5-fluoro-uracil (Champe and Benzer, 1962; Garen and Siddiqi, 1962), which acts by incorporation into the mRNA (Champe and Benzer, 1962) and by streptomycin (Whitfield et al., 1966), which

---

* r1589 is a deletion which removes a part of both the A- and B-genes and the "dividing-element" between them. A-function is lost in r1589, but B-function remains. It is suggested that the A- and B-gene fragments of r1589 are transcribed into a single messenger RNA, which is then translated into a single polypeptide chain having B-function (Benzer and Champe, 1962).

leads to misreading of the message (Davies *et al.*, 1965). When the phase of reading of a nonsense codon is shifted, by placing it between two phase-shift mutations of opposite sign, the effects of the nonsense mutation are no longer evident (Brenner and Stretton, 1965). In addition, it has now been shown *in vitro* that amber suppressors act at the level of translation of the genetic message (Capecchi and Gussin, 1965; Zinder *et al.*, 1966).

B. Structure of Amber and Ochre Codons

The ochre and amber codons have two nucleotides in common, since ochre mutants can be converted into amber mutants by a single mutation (Brenner *et al.*, 1965). (The conversion of a T_4 ochre mutant to an amber mutant can easily be detected by plating an ochre mutant lysate on an amber-suppressing strain and testing the revertants, which will include wild-types and amber mutants, on a suppressor-free strain, where the amber mutants will fail to grow.) By making use of the known specificities of transition mutagens, Brenner *et al.* (1965) have been able to deduce the structure of the two triplets.

Ochre mutants can be converted into amber mutants by a mutation that is strongly induced by 2-aminopurine. Since this mutagen induces the transition $A–T \rightleftharpoons G–C$ in both directions (Freese, 1959a, b), this means that one of the two triplets must contain a G–C pair. Ochre mutants cannot be induced to mutate to amber mutants with hydroxylamine, which produces unidirectional transitions of the $G–C \rightarrow A–T$ type (Freese *et al.*, 1961). This shows that it is the amber triplet which has the G–C pair and the ochre contains the A–T pair. The other two bases must be common to both triplets.

Both ochre and amber mutants can be induced by hydroxylamine from wild-type, proving that both triplets have at least one A–T pair. A further experiment with hydroxylamine as mutagen shows that the amber triplet has two A–T pairs and establishes the arrangement of the pairs with respect to the two strands of DNA. Hydroxylamine reacts only with cytosine in DNA, producing an altered base which behaves like thymine and therefore causes $G/C \rightarrow A/T$ transitions (Freese *et al.*, 1961). Since only one strand of DNA is transcribed into mRNA, it is possible to distinguish between mutants which require DNA replication in order to be expressed, and therefore result from a $C \rightarrow T$ change on the antisense strand, and those which are expressed immediately, and, therefore, are due to a change on the sense strand. Brenner *et al.* (1965) have measured the frequency of recurrence of different amber mutants induced by hydroxylamine after growth on two different bacterial hosts, one requiring the rII^+ phenotype, the other not. Two classes of amber mutants could be distinguished. Members of the first class occur with approximately equal frequency after growth on the two hosts, as expected for C→T changes on the antisense strand. Members of the second class were very rare after growth on the restricting host and must arise by C→T changes on the sense strand. This finding shows that the amber triplet must contain both A and U. The ochre triplet must also contain these bases, since the two triplets are connected by the single base changes already accounted for. These purely genetic experiments establish the amber and ochre triplets as (UAG) and (UAA) or (UAC) and (UAU), respectively.

The structure of the triplets has been more exactly defined by relating them to amino acids by mutation. In the T$_4$ head-protein system amber mutants are related to tryptophan and glutamine by transition* and to tyrosine by transversion. Amber mutants of the alkaline phosphatase system of *E. coli* have been related to leucine, serine, tyrosine, tryptophan, glutamine, lysine and glutamic acid and ochre mutants to all of these except tryptophan (Garen *et al.*, 1965). When these results are compared with the triplets defined by biochemical experiments, the unique assignment UAG can be made for the amber codon. It then follows that the ochre triplet must be UAA.

C. A Third Nonsense Codon, UGA

Weigert *et al.* (1966) have shown that the amber codon, UAG, is related to seven amino acids, including tryptophan, whereas the ochre codon, UAA, could be mutated only to six, tryptophan never being found. When an ochre codon was converted to an amber at the same site, the amber reverted quite readily to tryptophan, whereas the ochre codon could not be shown to give rise to tryptophan. These results strongly suggest that the codon UGA, which would be derived from UAA by a transition, does not code for an amino acid and, therefore, may be a nonsense codon. This evidence does not exclude the possibility that UGA may code for cysteine to which the codons UGU and UGC have been assigned.

Brenner *et al.* (1967), using the T$_4$*rIIB* cistron, have obtained a mutation which can give rise to the ochre codon (UAA) by mutagenesis with 2-amino purine. The change to ochre is also induced by hydroxylamine and does not require replication for expression. This suggests that the change arises from a G→A change in the mRNA. Only two codons can be related to UAA in this way, UGA and UAG.

---

* A *transition* is the replacement of a purine by another purine or a pyrimidine by another pyrimidine. *Transversion* involves the replacement of a purine by a pyrimidine or vice versa.

The mutant codon is not amber (UAG), since it is not suppressed by amber or ochre suppressors, and must therefore be UGA. Brenner *et al.* (1967) have also been able to produce UGA by selected phase-shifts early in the *rIIB* cistron. When (+ −) phase shifts are made over the first part of *rIIB*, the two phase-shift mutants frequently fail to suppress each other. This failure is not due to the generation of unacceptable amino acid sequences but to the generation of nonsense triplets ("barriers") in the shifted reading-phase. Amber and ochre codons have been identified as barriers. A third class of barriers can be mutated to the ochre codon by base-analogue mutagens and presumably consists of the UGA triplet. There is extensive evidence that the amino acid sequence coded by the first part of the *B* cistron is not critical for the function of the gene (Barnett *et al.*, 1967). This would suggest that the unacceptability of UGA does not result from the insertion of an amino acid, making it very likely that UGA is a nonsense codon.

A mutant in the *rIIA* cistron, *X665*, derived by mutagenesis with 2-aminopurine, maps identically with an amber mutant similarly derived. The only sense codons related to the amber codon by transition are CAG (glutamine) and UGG (tryptophan). The poor response of this amber mutant to the glutamine-inserting amber suppressor makes it likely that the amber mutant, and therefore also *X665*, was derived from a tryptophan codon (UGG). The mutant codon must then be either CGG (arginine) or UGA. This mutant, *X665*, when coupled with the deletion *r1589*, removes the *B*-activity of the resulting phage, and thus fulfils the original criterion of a nonsense mutant (Champe and Benzer, 1962).

This evidence does not prove formally that UGA does not code for cysteine, already assigned UGU and UGC. Two lines of evidence suggest that this is very unlikely. Genetic experiments with the T₄*rIIB* cistron suggest that UGA, like UAA and UAG, may be a chain-terminating codon. A mutation resulting in the reinitiation of polypeptide chain synthesis has recently been isolated in the *rIIB* cistron (Sarabhai and Brenner, 1967). This mutation requires the proximity of a chain-terminating codon (amber or ochre) in order to operate. Mutations of the UGA-type will substitute for amber or ochre in activating the chain initiation and are believed, therefore, to result in termination, like amber and ochre. Chemical evidence from Khorana's laboratory (Morgan *et al.*, 1967) supports the conclusion that UGA does not code for cysteine. Poly (UGA)ₙ used as the messenger in a cell-free system derived from *E. coli*, results in the formation of polymethionine (AUG) and polyaspartic acid (GAU). No other amino acids were incorporated and no polycysteine was found.

D. Nonsense Codons as Signals for Chain Termination

Termination of the growth of a polypeptide chain must involve the cessation of chain elongation and the cleavage of the ester bond between the carboxy-terminal residue of the chain and a tRNA molecule (Gilbert, 1963; Bretscher, 1963). There must be a special mechanism for this operation and a genetic signal which brings the mechanism into action. Since amber mutants have been shown, and the other nonsense mutants surmised, to result in chain termination, it may be supposed that amber, ochre or UGA provides this signal *in vivo*.

If a single triplet, such as the amber or ochre triplet, were the normal chain terminator, it would follow that chain termination would be suppressed in strains carrying nonsense-suppressor genes. The known amber suppressors, *suI*, *suII* and *suIII*, have efficiencies of suppression of about 50%, 30% and 60%, respectively (Garen et al., 1965; Kaplan et al., 1965), and yet these suppressors are not detrimental to the growth of the cell. This argues strongly against UAG being a common chain-terminating codon. The known ochre suppressors are all relatively weak, a finding which has made the ochre codon the most likely candidate as a chain-terminating signal. Both strong (60% efficiency) and weak (5% efficiency) suppressors of the UGA triplet have recently been isolated (Sambrook et al., 1967; Zipser, 1967).

It has been argued (Brenner, 1966) that chain termination may occur through the recognition of specific triplets by special tRNA molecules which do not carry amino acids. If this is the case, it might be possible to detect these molecules by looking for RNAs which decrease the efficiency of suppression in an *in vitro* suppression system.

E. Chain Termination in vitro

Capecchi (1967a) has described an assay for chain termination *in vitro*, using the RNA from an amber mutant of the RNA-phage R17 as messenger. The amber mutation, early in the coat-protein cistron, causes the release of an N-terminal hexapeptide fragment of the coat protein. The very rapid purification of this fragment forms the basis of the assay for chain termination. Using this assay, Capecchi (1967b) has isolated from *E. coli* a soluble protein which is required for chain termination. This release factor, which sediments at 4·2s, acts directly on the ribosome–mRNA–peptidyl tRNA complex. A requirement for a chain-terminating tRNA was sought but not found. The development of this assay and the isolation of the release factor represent important first steps in the understanding of the hitherto obscure biochemistry of chain termination.

IV. POLARITY

A. Translational Polarity

Certain mutations, in addition to eliminating the enzyme activity corresponding to the mutated gene, also lead to a marked reduction in the activities of the enzymes corresponding to some of the other genes in the same operon (Franklin and Luria, 1961; Jacob and Monod, 1961; Englesberg, 1961). This reduction is polarized, only those enzymes specified by genes between the mutated gene and the operator-distal end of the operon being affected (Franklin and Luria, 1961; Jacob and Monod, 1961; Ames and Hartman, 1963). This phenomenon has been termed "polarity". More recently it has been established that nonsense mutants of the amber (Weigert and Garen, 1965; Brenner *et al.*, 1965) and ochre class (Brenner and Beckwith, 1965) have polarity effects of this type (Newton *et al.*, 1965; Henning *et al.*, 1966). Phase-shift mutants are also polar (Martin *et al.*, 1966; Malamy, 1966; Imamoto *et al.*, 1966), due to the generation of nonsense codons within the altered reading frame (Martin, 1967).

Detailed studies of the effects of polar mutants on enzyme levels have been made using the *lac* operon of *E. coli* (Newton *et al.*, 1965), the *his* operon of *Salmonella typhimurium* (Fink and Martin, 1967; Martin *et al.*, 1966) and the *trp* operon of *E. coli* (Yanofsky and Ito, 1966). The conclusions to be drawn from each study are very similar and may be summarized as follows:

(1) Polar mutations primarily affect the synthesis of only those proteins specified by genes more distal from the operator than the gene containing the polar mutation.

(2) Nonsense mutations, but not missense mutations, are polar. Amber and ochre mutations at the same site are equivalent in the extent of their polarity.

(3) Within a single gene, the closer a nonsense mutation is to the operator-proximal end, the stronger will be its polar effect. This gradient of polarity appears to be more severe for the first gene of the operon.

(4) Suppressors of nonsense codons partially relieve the polar effect of a nonsense mutation.

Models have been proposed to account for these observations on polarity. In essence, these models suggest that ribosomes, having been stripped of the growing polypeptide chains at a nonsense codon (Sarabhai *et al.*, 1964), continue to move along the messenger but cannot translate the message until they encounter another chain-initiating signal at the beginning of the next gene. It is further supposed that ribosomes which are not involved in protein synthesis have an increased probability of becoming detached from the messenger RNA. Thus the longer the untranslatable region (i.e. the earlier the nonsense codon in a gene), the fewer the ribosomes reaching the next gene and the greater the polarity effect (Yanofsky et al., 1966; Martin et al., 1966). Superimposed on this model, Martin et al. (1966) have suggested that the degree of polarity may also be influenced by the efficiency of the particular initiation signal in rephasing the ribosome at the beginning of the next gene.

Malamy (1966) has observations on polarity in the lac operon of E. coli which lead him to suggest that a ribosome maintains a strict phase in an untranslatable region (i.e. proceeds along the mRNA three nucleotides at a time). Certain phase-shift mutations in the operator-distal ω-region of the lac z-gene are unusual in being completely polar. This complete polarity remains when the phase-shift mutations are preceded by less polar amber or ochre mutations in the z-gene. Thus the phase of the frame-shift mutations may still be recognizable in the untranslatable region following a nonsense codon. This interpretation must be viewed with caution, however, until these exceptionally polar phase-shift mutations are better understood.

B. Messenger RNA Levels in Polar Mutants

Polar mutants appear to terminate translation (Sarabhai et al., 1964) and their effects are suppressed by nonsense suppressors (Beckwith, 1963; Newton et al., 1965) which are known to act on the translation mechanism (Capecchi and Gussin, 1965; Zinder et al., 1966). Thus polarity appears to be the consequence of a translational defect. In contrast, Attardi et al. (1963) were unable to detect lac-specific mRNA in strong polar mutants of the lac operon of E. coli, so that here, at least, polarity appears to be affecting transcription.

A detailed study of the effects of polar mutations on specific mRNA levels has been made by Imamoto and Yanofsky (1967a, b) using the trp system of E. coli. This is a particularly suitable system for such a study because of the availability of non-defective transducing phages, φ80 pts, carrying various defined parts of the tryptophan operon (Matsushiro et al., 1965). Hybridization of pulse-labelled RNA from E. coli cells with the DNAs from a series of such phages can be used to detect intact trp-mRNA and also fragments of trp-mRNA corresponding to different regions of the trp operon (Imamoto et al., 1965a, b).

Using this technique, Imamoto and Yanofsky were able to show that polar mutations did affect the detectable mRNA levels. Strong polar mutations in the operator-proximal E-gene appeared to be defective in the production of that portion of the *trp*-mRNA molecule corresponding to the region of the operon beyond the site of the mutation. The rate of transcription of the region of the operon prior to the polar mutation appeared to be normal, but the majority of the *trp*-mRNA molecules produced were much smaller than the complete *trp*-mRNA. The size of the predominant *trp*-mRNA species increased in relation to the genetic distance from the beginning of the operon to the site of the mutation. Some full-length *trp*-mRNA molecules were produced by these mutants, in amounts correlating reasonably well with the extent of polarity.

The possibility remains, of course, that strong polar mutants produce full-length *trp*-mRNA molecules and the regions of the messenger that appear to be poorly translated are selectively degraded. Imamoto and Yanofsky (1967b) investigated this possibility by using increasingly short pulse periods. Varying the pulse period from 3 min to 40 sec failed to show any difference in the distribution of *trp*-mRNA in strong polar mutants. In addition, when the pulse was given during the first 6 min after derepression, reduced levels of *trp*-mRNA corresponding to the operon region beyond the nonsense codon were still detected. Since the growing mRNA chain is presumably still attached to the DNA at the site of synthesis during this period (Imamoto *et al.*, 1966), it seems unlikely that selective degradation of intact *trp*-mRNA molecules could be responsible for the short mRNA molecules detected.

Trp B- and *C*-gene mutations which show about 60% polarity on enzyme formation nevertheless produce normal *trp*-mRNAs. The observable polarity in these mutants, therefore, must be due solely to reduced translation and, conversely, reduced translation by itself does not necessarily lead to selective degradation of the messenger. In contrast, *E*-gene mutations of similar polarity markedly reduce the *trp*-mRNA levels.

These observations are probably best explained by the hypothesis that polarity is essentially caused by an untranslatable region in a polycistronic messenger, within which ribosomes are discharged at abnormal frequencies. In addition, the drastically reduced translation in some polar mutants leads to reduced transcription (Imamoto *et al.*, 1966; Imamoto and Yanofsky, 1967b). That, transcription and translation are closely coupled, has been previously proposed (Stent, 1964).

C. Polarity in vitro

The RNA phages f2 and R17 contain three cistrons, coding for the viral coat protein, the viral-specific RNA polymerase and a structural component, the A-protein (Gussin, 1966; Horiuchi et al., 1966). Non-sense mutations in the coat-protein cistron have been isolated and shown to exhibit a polar effect on the RNA polymerase cistron (Gussin, 1966; Lodish and Zinder, 1966). In both systems only nonsense mutations early in the coat-protein cistron have a measurable polarity (Gussin, 1966; Lodish and Zinder, 1966).

The RNA isolated from f2 and R17 can be used to stimulate *in vitro* protein-synthesizing systems derived from *E. coli* and phage coat protein can be identified in the product (Nathans et al., 1962). In addition, the coat proteins of these two bacteriophages do not contain histidine, so that the *in vitro* incorporation of this amino acid into protein, stimulated by phage RNA, can be considered a measure of the translation of non-coat genes. Then a comparison of histidine incorpora-tion stimulated by wild-type and coat-mutant RNA can be used as a measure of the effects of coat-protein cistron mutants on the translation of other phage genes.

When f2 or R17 containing an amber mutation very early in the head-protein cistron was used as message, the incorporation of histidine was suppressed to a few per cent of that obtained with wild-type RNA. This low incorporation was stimulated to approach wild-type levels by the addition of tRNA from cells containing an amber suppressor (Zinder et al., 1966).

The polar amber mutation in f2 RNA has further been shown to have an effect on polyribosome formation. When ^{32}P-labelled wild-type f2 RNA is added to a cell-free protein-synthesizing system from *E. coli* and the mixture is fractionated on a sucrose density gradient, the ^{32}P associates with a peak sedimenting at about 80s and also with two faster-sedimenting peaks. The 80s peak is presumed to be messenger RNA complexed with a single ribosome and the faster-migrating peaks to represent messenger RNA complexed with two or more ribosomes. When message containing the polar amber mutation is used, only monosomes are observed. The addition of suppressing tRNA to the system restores the presence of polysomes (Englehardt et al., 1967).

These observations on polarity *in vitro*, where the translation process is freed from dependence on transcription, confirm that polarity can be expressed at the translational level. The polysome patterns observed with the f2-amber mutant are consistent with the model suggesting discharge of ribosomes in an untranslatable region (Imamoto et al., 1966; Martin et al., 1966). Zinder et al. (1966) have suggested an alternative model which depends on unfolding of the message by ribosome movement during protein synthesis. This model proposes that the messenger must have several ribosome-binding sites and also some secondary structure which masks all but one of these sites. As protein synthesis proceeds the formerly masked binding sites become accessible, permitting further ribosome binding and the initiation of protein synthesis. The degree of polarity would then depend on how far along the messenger the ribosomes had travelled before reaching the chain-terminating codon.

D. Antipolarity

In studies of co-ordinate enzyme production in mutants of the *trp* operon of *E. coli*, Ito and Crawford (1965) noted that certain polar mutations caused a decrease in the amount of enzyme specified by the gene immediately preceding the mutated gene. This effect was termed "antipolarity" (Ito and Crawford, 1965). Yanofsky and Ito (1967) have extended these observations by using a set of well-defined ochre mutations in the *trp* A-gene to study the extent of antipolarity as a function of map locations. Each gene with a nonsense alteration was introduced into a strain carrying a mutation in the tryptophan repressor gene (R^- *trp*), so that enzyme levels could be measured under "gratuitous" conditions, obtained by growth in excess tryptophan.

Under these conditions ochre mutations throughout the A-gene had an antipolar effect on B-protein production. A gradient of antipolarity was evident, mutations nearer the B-gene having the greater effect on synthesis of B-protein. The effect was not limited to the B-gene, since indoleglycerol phosphate synthetase, specified by the C-gene, was also present in reduced amounts. The effect on the C-gene was less marked, however. Antipolarity does not appear to operate in the histidine operon of *Salmonella typhimurium* (Fink and Martin, 1967).

V. TRANSFER RNA AND SUPPRESSION

A. Nonsense Suppressors

Nonsense mutants of the amber and ochre type can be reverted by external suppressor mutations (Benzer and Champe, 1962; Garen and Siddiqi, 1962). The suppressors were divided into two classes on the basis of their specificity. Amber suppressors were defined by their ability to suppress amber mutants of T_4 (Epstein *et al.*, 1963). Ochre suppressors were defined by their ability to suppress amber mutants and also other nonsense mutants not suppressed by amber suppressors (Brenner and Beckwith, 1965). A third class of nonsense mutant has now been shown to be due to the triplet UGA (Brenner *et al.*, 1967) and suppressors for this class have been isolated (Sambrook *et al.*, 1967; Zipser, 1967).

Suppressor genes can be located on the genetic map by mating or transduction techniques (Signer *et al.*, 1965; Garen *et al.*, 1965). In such an experiment the donor strain usually carries the su^+-allele, the presence of which in the recombinants can be either scored or selected by its alteration of the mutant phenotype of a suppressible mutation in the recipient. The presence of suppressor genes in a bacterial strain can also be very conveniently scored by the ability of strains carrying the su^+-alleles to support the growth of suppressible mutants of phage (Benzer and Champe, 1962). The locations of several suppressor genes on the *E. coli* genetic map are given in Table 1.

Suppression has been shown to reverse the chain-terminating effect of amber codons by enabling the triplet to code for an amino acid and

TABLE 1
Characteristics of nonsense suppressors

Suppressor	Amino acid inserted	Efficiency of suppression	Map location*
Amber suppressors			
su_I^+	serine[5, 6, 7]	60%[3]	*his* (10%)[1, 2]
su_{II}^+	glutamine[3, 8]	30%[3]	*gal* (1%)[2]
su_{III}^+	tyrosine[3, 8]	50%[3]	*trp* (50%)[1]†
su_β^+	basic[4]	10%[4]	?
Ochre suppressors			
su^+B	?	5%[11]	*gal* (1%)[9]
$su^+C(su^+4)$	tyrosine	1%[11]	*trp* (50%)[9]†
su^+L	?	?	*gal* (55%)[12]
su^+M	?	?	*purD* (64%)[12]
su^+N	?	?	*aroC* (2%)[12]
su^+5[10] (su^+A[13])	basic[10]	3%[10, 13]	*gal* (10%)[10, 13]

1. Garen *et al.* (1965).
2. Signer *et al.* (1965).
3. Kaplan *et al.* (1965).
4. Kaplan, quoted in Stretton *et al.* (1966).
5. Stretton and Brenner (1965).
6. Weigert and Garen (1965).
7. Notani *et al.* (1965).
8. Weigert *et al.* (1965).
9. Beckwith (1963).
10. Gallucci and Garen (1966).
11. Brenner *et al.* (1966).
12. Eggerston (1968).
13. Yanofsky and Ito (1966).

* Figures in parentheses are % cotransduction of the su^+ and the given marker obtained with phage P1.

† Map order is: *cysB–trpEDCBA–ton*B–*Att*[80]–*gal*U–*tdk–su*$_{III}^+$–*su*$^+$*C* (Taylor and Trotter, 1967).

thereby allowing chain elongation to proceed (Sarabhai et al., 1964; Weigert and Garen, 1965; Stretton and Brenner, 1965; Notani et al., 1965). The amino acid inserted at the position specified by the non-sense codon can be determined by application of conventional techniques of protein chemistry to the protein produced by suppression. The amino acid inserted is characteristic of the suppressor gene. Thus the amber codon is translated as serine, glutamine and tyrosine in stocks carrying the su_I^+, su_{II}^+ and su_{III}^+ genes, respectively (Weigert and Garen, 1965; Stretton and Brenner, 1965; Weigert et al., 1965; Kaplan et al., 1965). The trp-linked ochre-suppressor gene, su_4^+, also leads to the insertion of tyrosine (Weigert et al., 1967) (Table 1).

Suppression by amber and ochre suppressors is incomplete. Stretton and Brenner (1965) have shown that when a T_4 amber mutant in the head-protein cistron of the phage is grown in a strain containing su_I^+, both the fragment characteristic of the mutant and the complete poly-peptide chain are made. The fraction of chains propagated is a direct measure of the efficiency of the suppressor. In the T_4 head-protein system, this can be determined by comparing the amounts of peptides C-terminal to the amber site with the amounts N-terminal to the site (Kaplan et al., 1965). In the alkaline phosphatase system of E. coli, Garen et al. (1965) determined efficiency of suppression by comparing the amount of immunologically cross-reacting material produced by a suppressed amber mutant and by the wild-type. This method gives a less reliable estimate since the measurements are necessarily made on different cultures which may not be in precisely the same state of derepression. The efficiencies of suppression are characteristic of the different suppressors. The three amber suppressors have efficiencies in the range 30–60% (Kaplan et al., 1965). Ochre suppressors are generally much less effective, having efficiencies in the range 3–15% (Gallucci and Garen, 1966; Yanofsky and Ito, 1966). Both strong (Sambrook et al., 1967) and weak (Zipser, 1967) suppressors for UGA have been isolated. The reason for the inefficiency of suppression is not yet understood. The level of suppression could be determined by competition between the suppressing tRNA and a tRNA which normally reads the nonsense codon (Kaplan et al., 1965).

The amino acids inserted by the known nonsense suppressors are all coded by triplets which are connected to the particular nonsense triplet by a single base-change. This is consistent with the idea that the suppressors are due to single base-mutations in the anticodons of the relevant tRNAs. Evidence on the in vitro suppression of phage amber mutants supports the idea that suppression in all three amber-suppressing strains is mediated by tRNA (Capecchi and Gussin, 1965;

Engelhardt *et al.*, 1965; Wilhelm, 1966; Gesteland *et al.*, 1967). If this hypothesis is correct, normal reading of amino acid triplets must be maintained. This would be possible if there were several gene-copies for each tRNA, or because of several ways of reading the same triplet (Crick, 1966).

Mapping studies show that the su_{III}^+ gene is closely linked to the attachment site for phage 80 (Signer *et al.*, 1965; Garen *et al.*, 1965). Thus defective transducing phages carrying su_{III}^+ can be prepared and, under appropriate conditions, used to increase greatly the number of copies of the suppressor-gene in the cell (Smith *et al.*, 1966). When lysogens carrying *80 dsu*$_{III}^+$ are induced, or when a suppressor-free strain is infected with *80 dsu*$_{III}^+$, lysis eventually results but can be prevented by the addition of chloramphenicol to the cells. Lysogenic strains incubated for 80 min after induction, then for a further 2 h in the presence of chloramphenicol, showed a 3·5-fold increase in the tyrosine-accepting tRNA relative to phenylalanine-accepting tRNA. Binding to ribosomes of tyrosine tRNA from *80 dsu*$_{III}^+$-infected cells was stimulated by UAU, UAC and UAG, whereas tyrosine tRNA from *80 dsu*$_{III}^-$-infected cells only responded to UAU and UAC. In cells infected with *80 dsu*$_{III}^+$, the amount of tyrosine tRNA binding with saturating levels of UAU and UAG was additive. This would suggest that the suppressor tRNA binds only with UAG and does not recognize the normal tyrosine codons. These experiments strongly suggest that the su_{III}^+ gene is a structural gene for a tyrosine tRNA. The relatively small increase in tyrosine tRNA synthesized after infection with the su^+-phage and the very small amount of UAG-recognizing tRNA normally present in su_{III}^+ cells suggest that the su-tRNA is only a minor fraction of the tyrosine tRNA of *E. coli* (Smith *et al.*, 1966). RNA–DNA hybridization experiments on the tRNA from cells infected with $\phi 80\ dsu_{III}^+$ support this conclusion (Landy *et al.*, 1967).

B. Missense Suppressors

Missense mutants, as well as nonsense mutants, can be reverted by external suppression. Although missense suppression has been observed in several bacterial systems, the only well-characterized ones are those arising as suppressors of certain missense mutants in the tryptophan synthetase *A*-gene of *E. coli*.

Tryptophan synthetase *A*-gene mutant *A23* has a replacement of a wild-type glycine residue by arginine (Helinski and Yanofsky, 1962). One of the revertants of *A23* is due to an external suppressor mutation, su_{23}^+(Brody and Yanofsky, 1963). In strains carrying *A23* and su_{23}^+, two structurally distinct A-proteins are produced. The major component is the mutant enzyme and a minor component has the wild-type sequence restored (Brody and Yanofsky, 1963). Mutant $A78$ has a wild-type glycine residue replaced by cysteine (Guest and Yanofsky, 1965). Suppression of this mutation by the external suppressor gene su_{78}^+ results in the production of a wild-type like A-protein but the primary structural change in this protein has not been established. A third suppressor, su_{58}^+, reverts the phenotype of mutant $A58$, which has an aspartic acid residue in place of a wild-type glycine residue (Guest and Yanofsky, 1965). The effect of su_{58}^+ is not specific to $A58$, since the suppressor also phenotypically reverts another mutant which contains the same amino acid replacement as $A58$ but at a different location (Berger and Yanofsky, 1967).

It is evident from these observations that missense suppression can be considered a mistake in the translation process, involving misreading of a specific codon. Further information on the mechanism of suppression by su_{23}^+ and su_{78}^+ has been obtained from *in vitro* studies using synthetic RNAs containing alternating sequences of two bases as messenger. Alternating poly AG normally directs the synthesis of an alternating arginine-glutamic acid polypeptide in a cell-free system from *E. coli* (Jones *et al.*, 1966). The use of tRNA from cells carrying the su_{23}^+ mutation led to the incorporation of glycine, as well as arginine and glutamic acid, into polypeptide (Carbon *et al.*, 1966). Similarly, poly UG, which normally directs the synthesis of an alternating valine-cysteine polypeptide, led to incorporation of glycine in the presence of tRNA from cells carrying su_{78}^+ (Gupta and Khorana, 1966). Thus in both these examples of missense suppression, as is the case for nonsense suppression, suppression is caused by a genetically altered tRNA. Which tRNA species is involved, and the exact nature of the mutational alteration, remain to be demonstrated.

No example has yet been found of suppression mediated by an altered activating enzyme. The tRNA-suppressors are presumably relatively common because of the possible existence of multiple gene-copies for many tRNAs, and of reading overlap due to multiple codon–anticodon pairing possibilities ("wobble": Crick, 1966). If a mutational event completely changed the specificity of charging of an activating enzyme, the mutation would probably be lethal. There seems no *a priori* reason why a mutant activating enzyme might not produce a low level of charging of the "wrong" tRNA, while still retaining its original charging function, and thus produce suppression. It might be profitable to look for this type of suppressor in strains which have been made diploid for various large regions of the chromosome and therefore carry at least two copies for many genes.

It has recently been shown that the haemoglobin α-chains from both rabbit (von Ehrenstein, 1966) and mouse (Rifkin *et al.*, 1966) contain alternative amino acid residues at certain positions in the polypeptide chain. It is suggested that these ambiguities may be due to ambiguous translation, analogous to missense suppression, although in neither case has the possibility of heterozygosity been eliminated.

VI. INITIATION OF TRANSLATION

The biochemical study of polypeptide chain initiation received great impetus from the discovery by Marcker and Sanger (1964) of a species of methionyl-tRNA which can be formylated to give N-formyl-methionyl-tRNA. The recent developments stemming from this discovery have been covered by Bretscher and Jones (1967) in Vol. I of this series and will not be dealt with here. Suffice it to recall that N-formyl methionyl-tRNA$_F$ acts as a chain-initiating RNA, irrespective of the presence or absence of the formyl group, because of its high specificity for the "peptide site" on the ribosome, when all other charged tRNAs bind to the "amino acid site". The binding of N-formyl methionyl-tRNA$_F$ to ribosomes is stimulated by the nucleotide triplets AUG, UUG, and GUG codons assigned to methionine, leucine and valine, respectively (Ghosh *et al.*, 1967). In addition, synthetic messengers with AUG, GUG or GUA near the 5′ end initiate efficiently and show N-formyl methionyl-tRNA$_F$-dependent polypeptide synthesis (Ghosh *et al.*, 1967). For these reasons it is believed that AUG, GUG and GUA all may act as initiation signals *in vivo*.

It has recently been shown that N-formyl methionyl-tRNA$_F$ differs from other tRNAs in binding only to the 30s ribosomal subunit in the presence of natural mRNA, when other tRNAs bind only to the 70s ribosome (Nomura and Lowry, 1967). It now seems likely that 70s ribosomes may not exist free in the cell but are formed from the 50s and 30s components on initiation of translation and dissociated to these components on chain termination (Schlessinger *et al.*, 1967). This suggests that the absence or presence of 70s ribosomes determines whether a codon acts as an initiator or is merely translated into an amino acid.

The geneticists' contribution to the study of chain initiation has so far been very limited. Recently, however, Sarabhai and Brenner (1967) have isolated a mutant of T$_4$ which is able to reinitiate polypeptide chain synthesis after chain termination by nonsense codons. A double mutant, *360 + FCO*, containing an ochre codon (*360*) and a phase-shift mutation (*FCO*) in the *rIIB* cistron, was induced to revert by 2-amino-

purine, a mutagen which causes transitions (Freese, 1959a, b). This
is a surprising result, since a phase-shift mutation is only expected to
revert by the introduction of a second compensating phase-shift event
(Crick et al., 1961). The induced revertants produced only minute
plaques on a suppressor-free K12 (λ) host, and a typical r-plaque on
E. coli B.

The nature of the new mutation, x, was analysed by crossing the
minute revertant, $360 + x + FCO$, with wild-type phage and isolating
segregants unable to grow on su^-K12. In this way it was shown that x
itself was not mutant and that the (+) phase-shift was essential for the
minute phenotype on the su^-K12 host. Further crosses established
that FCO could be replaced by any of a series of (+) phase-shift
mutations to the right of x, but not by (−) phase-shift mutations. The
map order of the mutations in the minute revertant was shown to be
$360 - x - FCO$, meaning that x could not be phasing-out the ochre
mutation, 360.

To explain these observations, Sarabhai and Brenner suggest that x
provides a mechanism for reinitiation of the polypeptide chain after
termination by a nonsense codon. The new starter must be in a (−)
phase, because the presence of a (+) phase-shift to the right is obligatory. x is not a (−) phase-shift mutation because it has no phenotypic
effects by itself. This hypothesis demands that a polypeptide fragment,
extending from x to the end of the B-gene and containing a phase-
shifted sequence between x and FCO, be functionally active. There is,
however, considerable evidence suggesting that the first part of the
B cistron is not necessary for function (Champe and Benzer, 1962;
Barnett et al., 1967). Indirect evidence was presented to support the
suggestion that two distinct polypeptide chains are produced in the
triple mutant.

By constructing stocks containing the appropriate combinations of
genetic markers, Sarabhai and Brenner were further able to show that
the obligatory chain termination could occur on either side of the
initiation mutation x but that it had to be quite close to it (within
about 10 nucleotides). In addition, restarting was independent of the
phase between the terminator and starter, since phase-shift mutations
inserted between the terminator and starter were without effect.

It will be difficult to determine the sequence of bases which give rise
to the reinitiation phenomenon because of the absence of a chemically
analysable gene-product for the $rIIB$ system. A genetic analysis using
mutagen specificity, so effectively used to determine the structure of
the amber and ochre codons (Brenner et al., 1965), might further
elucidate the nature of the initiation sequence.

VII. MUTATIONS AFFECTING RIBOSOMES

Genetic information about the ribosome is still very limited, due largely to the lack of a suitable method for the isolation of mutants affecting the organelle. In addition, even when a mutation is known to affect the activity of the ribosomes *in vitro*, it can be extremely difficult to decide if the mutation is in a structural gene for a ribosomal component, or in a gene specifying an enzyme which might affect the ribosome indirectly.

One approach to the problem involves the selection of mutations conferring resistance to antibiotics known to act by interaction with the ribosome. Streptomycin, for example, which is believed to act by interaction with the 30s ribosomal subunit (Speyer *et al.*, 1962; Flaks *et al.*, 1962; Davies *et al.*, 1964; Cox *et al.*, 1964), gives rise to resistant mutants of *E. coli* which have an alteration of the 30s subunit (Staehelin and Meselson, 1966; Traub *et al.*, 1966). Apirion (1967) has used erythromycin and lincomycin, both of which affect the 50s ribosomal subunit (Taubman *et al.*, 1966; Chang *et al.*, 1966), to select mutants affecting ribosomal structure. One mutation to increased lincomycin sensitivity was closely linked to the *str* locus. Mutations at three other genetic loci were also obtained. That these loci specify products affecting the ribosome was confirmed by showing differences from the parental strain in polypeptide formation *in vitro* on ribosomes from strains of each mutant class (Apirion, 1967). Resistance to spectinomycin, which is expressed in the structure of the 30s subunit, is genetically linked to the *str* locus (Davies *et al.*, 1965).

A second approach to the genetics of the ribosome has been the isolation of temperature-sensitive mutations. Flaks *et al.* (1966) have specifically screened for temperature-sensitive mutations which are linked to the *str* locus. Two such mutants, mapping between the *str* and *sp* loci, showed temperature sensitivity of *in vitro* protein synthesis. No significant temperature sensitivity of the ribosomes was found, however, so that the hypothesis that these mutations represent temperature-sensitive alterations in ribosome function remains unproved.

Streptomycin, which inhibits protein synthesis by the interaction with the 30s ribosomal component, also mimics suppressor mutations in increasing the ambiguity of code translation (Gorini and Kataja, 1964; Davies *et al.*, 1964). This suggested the possibility that one class of suppressor mutations might affect the structure of the ribosome. Apirion (1966) has isolated temperature-sensitive revertants of an *E. coli* tryptophan auxotroph and shown the reversion to be due to a suppressor mutation. Transduction experiments suggested that temperature sensitivity and suppressibility were due to the same mutation.

3

The ribosomes of the suppressible revertant were shown to be more temperature-sensitive than those of the parent strain in their capacity to synthesize polyphenylalanine, and the 50s subunit was the temperature-sensitive component. No attempt was made to prove that the altered ribosomal behaviour and the suppression were due to the same mutational event. This is a serious omission, since the suppressed mutants were isolated following treatment with N-methyl-N'-nitro-N-nitrosoguanidine (NNG). There are many examples of multiple mutational events caused by NNG. Flaks *et al.* (1966), for example, were able to show three closely linked mutations following a single treatment with this mutagen.

MacDonald *et al.* (1967) have isolated a mutant of *E. coli* in which the maturation of the 50s ribosomal subunit appears to be affected. The mutant was isolated on the basis of its abnormally high ratio of RNA to protein which increases its buoyant density in a caesium sulphate density-gradient. The strain contains an elevated level of a component which sediments at about 43s, which is believed to be a direct precursor of the 50s ribosomal subunit. Isolated 70s ribosomes from the mutant strain are less efficient than those of the parent strain in supporting protein synthesis *in vitro*. Such a mutant should be very useful for studying the composition and maturation of the larger ribosomal subunit.

VIII. REGULATION OF PROTEIN BIOSYNTHESIS

A. The Repressor/Operator Model

The celebrated Jacob and Monod (1961) theory of the regulation of gene activity was largely based on observations with the lactose operon of *E. coli*. The lactose operon consists of three adjacent genes, the z, y and a genes, coding for β-galactosidase, the β-galactoside permease and β-galactoside transacetylase, respectively. The function of the latter enzyme is unknown and the a-gene can be deleted without affecting growth on lactose (Fox *et al.*, 1966). In wild-type strains of *E. coli* grown on almost any carbon source, the activities of these genes are repressed, the corresponding enzymes being detected in very low amounts. Growth on lactose as sole carbon source, or the addition to the growth medium of various compounds structurally related to lactose, results in the induction of gene expression and an increase in the amounts of the *lac* proteins of up to 1000 times.

Much of the evidence for the Jacob–Monod model has come from constitutive mutants, which form high levels of the *lac* enzymes in the absence of inducer. (Constitutive mutants for inducible systems can often be selected by using as sole carbon or nitrogen source a compound which is a substrate of the enzyme system, but not an inducer of its synthesis.) Such mutations invariably affect simultaneously the levels of all three proteins of the *lac* operon.

Constitutive mutants of the *lac* operon fall into two classes on the basis of both map location and behaviour in heterozygotes. One class, i^- mutants, maps in a region separate from but closely linked to the *lac* operon (see Fig. 10). Heterozygotes of structure $i^+z^-/F'i^-z^+$ are phenotypically inducible, indicating that the inducible allele (i^+) is dominant over the constitutive and that it is active in the *trans* position. The second type of constitutive mutant, termed operator-constitutive or o^c, maps in the *lac* region, between i and z. [This type of mutation is relatively rare compared with the i^- type and was originally isolated in cells diploid for the *lac* region which virtually eliminates the recessive (i^-) constitutive mutants (Jacob *et al.*, 1960).] Cells of composition $o^+z^+/F'o^cz^+$ are constitutive, while $o^+z^+/F'o^cz^-$ are inducible, indicating that the o^c mutation has no effect in the *trans* position. A second type of i-gene mutation, i^s, results in a *lac⁻* phenotype and is *trans* dominant to i^+ but recessive to o^c (Willson *et al.*, 1964).

The Jacob–Monod model explaining these phenomena is shown schematically in Fig. 10. The three genes of the *lac* operon are transcribed into a single polycistronic messenger RNA. The i-gene is the structural gene for the *lac*-specific repressor molecule; the o^c mutants define the operator region, the site of action of the *lac* repressor. In an uninduced wild-type strain, the repressor binds to the operator-site and thus prevents transcription of the operon into messenger RNA. The presence of an inducer removes the repressor from the operator, allowing transcription to proceed. i^- mutations result in an inactive repressor and i^s mutations in a repressor which cannot be antagonized by the inducer. In support of this hypothesis, it has been shown by DNA/RNA hybridization studies that the level of operon-specific mRNA in the cell is greatly increased on induction or in a constitutive mutant (Attardi *et al.*, 1963). Indirect evidence supports the hypothesis that the information from the structural genes of the *lac* operon is contained in a single mRNA molecule. Polar mutants in the z-gene abolish or greatly reduce the expression of the y- and a-genes (Franklin and Luria, 1961; Jacob and Monod, 1961). Since these mutants are believed to exert their effects on translation, the permease and trans-acetylase must be translated from the same mRNA as β-galactosidase. The observation that polar mutants reduce transcription (Contesse *et al.*, 1966) weakens this argument. Kiho and Rich (1966) have shown that an amber mutation in the y-gene affects the size of the polysome on which β-galactosidase is made. If the y- and z-genes were transcribed into different mRNAs, no effect should have been observed.

Biosynthetic enzymes are usually subject to repression in the presence of the end-product of the pathway. This type of control can easily be fitted to the Jacob–Monod scheme by assuming that the

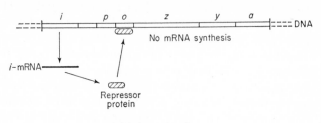

(a) Inducer absent

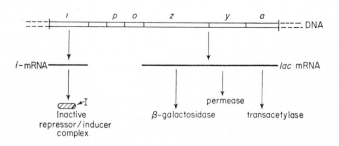

(b) Inducer present

Fig. 10. The Jacob–Monod model for the control of expression of the *lac* operon of *E. coli*. i is the structural gene for the repressor protein, p the promoter region, o the operator and z, y and a the structural genes for β-galactosidase, permease and transacetylase, respectively. The location of p on the map is due to Ippen *et al.* (1968). In the absence of inducer, the repressor binds to the DNA of the operator region and prevents transcription. In the presence of inducer, the repressor protein is removed from the operator, allowing synthesis of the *lac* mRNA to proceed (after Jacob and Monod, 1961).

repressor can only bind to the operator in the presence of the end-product (or a co-repressor derived from the end-product). In the absence of co-repressor the repressor is removed from the operator, allowing transcription to proceed. Constitutive mutants of repressible systems can be isolated, usually by selecting for resistance to a non-metabolizable analogue of the specific end-product. Mutants having the properties of both regulator-gene and operator constitutives have been isolated for several repressible systems (e.g. Matsushiro *et al.*, 1965; Roth *et al.*, 1966; Margolin and Bauerle, 1966).

B. The Nature of the Repressor

The high structural specificity of the interaction of the inducer (or co-repressor) with the repressor strongly suggested that the repressor must at least contain a protein. There is both genetic and chemical evidence that the i-gene product is a protein. The genetic evidence is the existence of i^- mutations which are sensitive to amber suppression (Bourgeois et al., 1965; Muller-Hill, 1966). Since it is known that such mutants affect translation, the i-gene messenger must be translated into protein. The chemical evidence depends on the isolation and partial purification of the i-gene product, using the binding of the non-substrate inducer, isopropyl-β-D-thiogalactoside (IPTG) measured by equilibrium dialysis against radioactive IPTG, as an assay (Gilbert and Muller-Hill, 1966). In this way a protein was isolated which had a binding constant of $1 \cdot 3 \times 10^{-6}$ M for IPTG, and sedimented at between 7s and 8s. No such binding could be detected with extracts from strains carrying an amber mutation in the i-gene, a deletion of the i-gene or an i^s mutation. The protein was estimated to occur in about 10 copies per gene and its concentration in the cell did not vary on induction of the lac operon. By purifying ^{35}S-labelled repressor, Gilbert and Muller-Hill (1967) have been able to show that the lac repressor binds specifically to lac-DNA and is removed by IPTG. Lac-DNA carrying an o^c mutation shows greatly reduced binding. The specific interaction is unaffected by RNase, is salt-sensitive and is eliminated by prior denaturation of the DNA.

Ptashne has applied the extensive knowledge of the genetics and physiology of λ-infection to obtain the most favourable conditions for the specific labelling and subsequent isolation of the phage λ repressor (Ptashne, 1967a). Ptashne's work is an excellent illustration of the way in which genetic techniques can be used to facilitate and enrich the study of biochemical systems. A strain of E. coli lysogenic for λ ind^-, which is not induced by UV light, was subjected to massive doses of UV irradiation in order to destroy host protein synthesis. A portion of the cells was infected at a multiplicity of 30–35 with λ-phages carrying mutations in gene N, and a second portion infected with λ carrying an additional mutation in the CI-gene and therefore making no repressor. [The mutations in gene N prevent the production of many λ-specific proteins which would otherwise be synthesized after infection at these multiplicities.] One culture was then labelled with [3H]leucine, the other with [14C]leucine. The cells were mixed and sonicated and the extract fractionated to look for a protein marked with one label but not the other. A single differentially labelled protein peak was isolated following chromatography of the supernatant fraction on DEAE-cellulose. This protein was missing from cells infected with phages carrying amber mutations in the CI-gene and was made in modified form by phages having temperature-sensitive mutations in the CI-gene. The repressor-protein was shown to be acidic and to have a molecular weight of approximately 30,000.

Sucrose gradient sedimentation experiments demonstrated that the isolated CI-gene product bound specifically to native λ-DNA (Ptashne, 1967b). It did not bind to DNA from phage λ imm^{434}, which carries the gene specifying the 434 repressor, or to denatured λ-DNA. The binding was very sensitive to salt concentration and may therefore be at least partly electrostatic. Ptashne estimated a repressor-operator affinity of 10^{-9} to 10^{-10} M.

The finding in two different systems that an isolated repressor binds specifically and with high affinity to the relevant DNA strongly suggests that the repressor blocks transcription *in vivo* by direct binding to DNA.

C. The Promoter

Genetic evidence has suggested the existence of another function necessary for expression of the *lac* operon in *E. coli*. Starting with a homozygous i^s diploid, it is possible to select for o^c mutations that restore expression of the *lac*-genes (Jacob et al., 1964). By selecting for o^c mutations at 42° on melibiose as carbon source, under which conditions only *y*-gene function is required,* o^c mutants are found which are $z^- y^+$ (Beckwith, 1963; Jacob et al., 1964). Of 67 o^c mutations of this type, *all* were shown to have lost *i*-gene activity, presumably due to extensive deletion. In contrast, o^c mutations which retained z^+ activity ended either between o and i or within i (Jacob et al., 1964). On this basis it was suggested that deletions bridging the $o-z$ boundary removed a function, termed the "promoter", which is essential for expression of the *lac* operon. These deletions must fuse the intact *lac*-, *y*- and *a*-genes to other operons, which in two cases have been identified (Jacob et al., 1965; Beckwith et al., 1966). In these cases the intact *lac*-genes are under the control system of the other operon and have altered maximum levels of expression (Jacob et al., 1965; Beckwith et al., 1966).

These original observations on the promoter now appear to be

* In *E. coli* melibiose, an α-galactoside, is not a substrate for β-galactosidase. At 42°, however, its transport into the cell is mediated by the β-galactoside permease, making the metabolism of melibiose dependent on a functional *y*-gene.

unsound, since Davies and Jacob (1968) have recently shown that the apparent $i^-z^-y^+$ "revertants" are not overlapping deletions; they are double mutations, one in the i-gene and one in the z-gene.*

Scaife and Beckwith (1966) have isolated leaky lac^- mutants of *E. coli* which have a much reduced maximum level of gene expression. These mutations are *cis*-dominant, affect all three lac-genes simultaneously and are independent of operon regulation by repressor–operator interaction. It is suggested that these mutants contain a mutation in the promoter site, as a result of which the rate of initiation of operon transcription has been reduced (Scaife and Beckwith, 1966). By mapping against a series of deletions entering the operon from the a-gene end, the leaky mutants have been located between i and o (Ippen *et al.*, 1968). This suggests the possibility that the repressor might act by directly blocking the progress of the RNA polymerase into the structural genes of the lac operon (Ippen *et al.*, 1968). The question of whether or not the operator region is transcribed remains unanswered.

Mutations having promoter-like properties have been isolated in the trp operon of *Salmonella typhimurium* (Margolin and Bauerle, 1966). The five genes of the trp operon can be divided into two groups with respect to expression (Margolin and Mukai, 1964; Margolin, 1965). Deletions in the region of the trp operon can be isolated by selecting for reversion of a polar leucine mutation. Amongst these revertants are a class in which the $supX$ locus, closely linked to trp, has been deleted. Some of the deletions extend into the trp operon (Mukai and Margolin, 1963).

All deletions that terminate before $trpE$† result in non-function of $trpA$ and $trpB$ but continued expression of $trpE$, $trpD$ and $trpC$. The latter three genes function constitutively, producing enzymes at about 50% of the fully repressed wild-type level. Deletions terminating in $trpE$ or $trpD$ completely abolish function of the remaining intact genes. Strains harbouring deletions entering the A or B genes, and therefore failing to express B-gene function, can be reverted to do so by selecting for growth on anthranilic acid. Similarly, strains carrying deletions entering the E-gene, and therefore failing to express D-gene function, can be reverted by selecting for growth on indole. Such revertants

---

* It may be recalled that the melibiose-fermenting "revertants" were selected from an i^s diploid strain: thus a typical i^- mutation should not lead to expression of the lac operon, since i^s is dominant over i^-. The particular i^- mutations isolated here are unusual, however, in being *trans*-dominant over i^s (Davies and Jacob, 1968).

† Note that corresponding genes of the trp operons of *E. coli* and *S. typhimurium* have different designations: *E. coli* O-EDCBA corresponds to *S. typhimurium* O-ABEDC!

occur spontaneously or can be induced by base-substitution mutagens but not by frame-shift mutagens. It is suggested that the reversion events are single base-pair substitutions resulting in the formation of nucleotide sequences that serve to initiate gene expression (Margolin and Bauerle, 1966).

D. Positive Control Systems

The essential feature of the Jacob–Monod theory of regulation of gene expression is that the control is negative with the controlling element, the repressor, "switching off" gene expression. This type of control is apparently not universal and there are now several good examples of positive-control systems.

The best-studied positive-control system is the L-arabinose operon of *E. coli*. The L-arabinose gene-complex consists of genes D, A, B and C, linked in that order between the markers *thr* and *leu*, and an unlinked gene E. Genes D, A and B are structural genes for three inducible enzymes involved in the conversion of L-arabinose to D-xylulose-5-phosphate, and gene E the structural gene for L-arabinose permease. Arabinose-negative mutations have been isolated which map in the C-gene and which lead to the inability to induce the synthesis of the enzymes of the L-arabinose operon (Englesberg, 1961) and the L-arabinose permease (Novotny and Englesberg, 1964). Complementation studies have shown that the C-gene is separate and distinct from genes A, B and D. In addition, the C^- allele is recessive to C^+ (Helling and Weinberg, 1963; Englesberg *et al.*, 1965). Constitutive mutants (C^c), mapping in the C-gene, are *trans*-dominant to C^- but *trans*-recessive to C^+ (Englesberg *et al.*, 1965; Sheppard and Englesberg, 1966).

The behaviour of these C-gene regulatory mutants in dominance tests is clearly different from that of the known regulatory mutations of the *lac* operon of *E. coli*. The observations are all consistent with a model in which the C-gene product is required for expression of the operon but is only active in the presence of the inducer, L-arabinose. On this model the C^c-gene product is active without the intervention of inducer. The dominance of C^+ over C^c can be explained by assuming that the C-gene product is a repressor which is converted to an activator on interaction with the inducer. Then a C^c/C^+ diploid would remain repressed by the C^+-gene product until induced by L-arabinose. Alternatively, the activator could be a multimeric component, with the activity of the C^c-gene product being neutralized by combination with the C^+-gene product.

The behaviour of regulator-gene deletions in the lactose and arabinose systems of *E. coli* illustrates the contrast between these two systems. *Lac i*-gene deletions produce a constitutive phenotype (Beckwith *et al.*, 1966), whereas arabinose *C*-gene deletions are arabinose-negative (Sheppard and Englesberg, 1967). This observation strongly supports the interpretation that the *C*-gene product regulates the arabinose operon in a positive manner.

The formal distinction between positive and negative control is very slight indeed and to classify a system as one or the other requires a precise knowledge of its mechanism. In this regard the isolation and *in vitro* characterization of the product of the arabinose *C*-gene will be of great interest.

Other systems in which there is preliminary evidence for positive control include the galactose enzymes in *Saccharomyces* (Douglas and Hawthorne, 1964; 1966) and the enzymes for L-rhamnose utilization in *E. coli* (Power, 1967).

The control of development in phage λ provides an example of the interaction of positive and negative control processes. Defective mutants of several cistrons in λ give a burst after infection of a bacterial strain lysogenic for a closely related but hetero-immune bacteriophage (Thomas, 1966). There is no phage production after infection of the corresponding non-lysogenic strain. This effect, "prophage complementation", results from induction of prophage genes following the hetero-immune superinfection. [In some cases, but by no means all, it has been shown that the complementation is not simply due to recombination between the prophage and the superinfecting phage (Thomas, 1966).] This means that the control exerted by the prophage repressor can be by-passed in a *trans* position, suggesting that these particular prophage genes are not blocked by the repressor. It is suggested that the λ-repressor prevents the synthesis of a diffusable substance which normally acts as an inducer of these genes. By determining which genes are expressed after hetero-immune superinfection, it was concluded by Thomas (1966) that gene *N* has the key role in producing the early inducer and is itself under the control of the phage repressor. It has been suggested that gene *Q* controls the induction of the late functions of phage (Dove, 1966; Joyner *et al.*, 1966). Thomas' results suggest that gene *Q* is itself induced by the inducer of the early functions (Thomas, 1966). The regulatory relationships in λ, which are beginning to be understood in some detail, may serve as a simple model for the general mechanisms of the ordered control of gene expression in developmental processes.

E. Catabolite Repression

The inducible synthesis of many catabolic enzymes is influenced by the nutritional condition of the cells. The differential rates of induced enzyme synthesis are much reduced in media containing rapidly metabolized carbon sources, despite the presence of high concentrations of inducer (Cohn and Horibata, 1959). Reduction in biosynthetic activity, produced, for example, by nitrogen starvation or absence of a required amino acid, greatly enhances this repression (Mandelstam, 1961). It is suggested that the carbon sources give rise to a high concentration of catabolites, one of which brings about a specific repression. This phenomenon has been called catabolite repression (Magasanik, 1961).

McFall and Mandelstam (1963a, b) showed that the glucose-mediated repression of β-galactosidase, serine deaminase and tryptophanase in $E.$ $coli$ is due not to glucose itself but to metabolites of glucose which are intermediates in the degradative pathways of the substrates of these enzymes. This suggested that catabolite repression is not a general repression effect but is mediated by a specific catabolite for each system.

Differential pulse-labelling experiments with the lac system of $E.$ $coli$ have shown that catabolite repression limits transcription of the operon (Attardi et $al.$, 1963). Thus, since both induction and repression are operon-specific, and both affect the synthesis of messenger RNA, the possibility was considered that the two phenomena operate via the same regulation system (i.e. the repressor/operator system). Since it has been shown that lac i^- mutants of $E.$ $coli$ remain susceptible to catabolite repression (McFall and Mandelstam, 1963; Loomis and Magasanik, 1964), it was concluded that the repression could not operate via the i-gene product. This conclusion has been doubted on the grounds that many i-gene point mutations will produce an inactive i-gene product, which could nevertheless be activated by interaction with a catabolite co-repressor (Clarke and Brammar, 1964; Palmer and Moses, 1967). Palmer and Moses (1967) have studied what they term "transient catabolite repression" in strains carrying an amber mutation or a temperature-sensitive mutation in the lac i-gene. In both cases the repression was abolished under non-permissive conditions, strongly suggesting the involvement of the i-gene product in this repression phenomenon.

Conflicting reports exist on the involvement of the $E.$ $coli$ lac-operator region in catabolite repression. Loomis and Magasanik (1964) showed that a lac o^c mutant was still catabolite-repressible, whereas Palmer and Moses (1967) show that another lac o^c mutation abolished this repression. A possible reason for this apparent dichotomy is that the two groups may be studying different phenomena, since the effect studied by Palmer and Moses appears to differ from that studied by Magasanik's group in being essentially transient.

Methods have been described for isolating mutants which are resistant to catabolite repression. Loomis and Magasanik (1965) obtained such mutants for the *E. coli lac* operon by using a poor substrate (N-acetyl lactosamine) as sole nitrogen source in the presence of a nonsubstrate inducer and 0·4% glucose. Zwaig and Lin (1965) described an autoradiographic method for the detection of colonies with altered catabolite repressibility. The method depends on the resistant mutant's having a relatively rapid incorporation of label from a ^{14}C-labelled substrate in the presence of glucose. When isolating such mutants by any technique it is essential to show that the mutant phenotype is not merely due to disturbed metabolism of the carbon source.

A mutation (CR^-), conferring insensitivity to catabolite repression of the *lac* operon of *E. coli*, was shown to map genetically distant from the *lac*-genes (Loomis and Magasanik, 1965). The construction of stable merodiploids heterozygous for the CR-gene showed the wild-type allele to be *trans*-dominant (Loomis and Magasanik, 1967). It was therefore suggested that the CR-gene determines a cytoplasmic product which, when activated by the catabolite co-repressor, acts as a repressor of the *lac* operon (Loomis and Magasanik, 1967).

F. Genetic Modification of Amino Acid-activating Enzymes

The genetic approach to the study of amino acid-activating enzymes is largely dependent on the use of "conditional" mutations, which exert their phenotypic effect under one environmental condition ("restrictive condition"), but not under another ("permissive condition"). The study of mutant activating enzymes has been largely confined to temperature-conditional mutations.

1. Isolation of Mutant Activating Enzymes

The technique for the isolation of temperature-conditional amino acid-activating enzyme mutants has been outlined by Neidhardt (1966). A bacterial culture is mutagenized and mutants are isolated which fail to grow under the restricting conditions (say 42°) but will grow under other conditions (say 35°). Penicillin treatment under restrictive conditions can be used to increase the relative frequency of the conditional mutants. The specific enzyme lesion in the mutants must then be

identified. Conditional mutants are shifted from permissive to restrictive conditions and the kinetics of RNA, DNA and protein synthesis are
followed. Mutants suspected of having primary lesions in protein
synthesis are then surveyed for aminoacyl-tRNA synthetase activities.
When mutagens have been used it is essential to establish that the
enzyme alteration detected *in vitro* is the cause of the changed physiology. This can be done by showing that both the ability to grow under
the restricting conditions and normal enzyme activity are restored by
a single mutational event. This is best achieved by plating under
restricting conditions for spontaneous revertants, or by transduction
with phage grown on a wild-type host (Neidhardt, 1966).

Non-conditional mutants with altered aminoacyl-tRNA synthetases
have been isolated by selection for resistance to amino acid analogues.
Thus a p-fluoro-phenylalanine-resistant mutant of *E. coli* was shown to
have a phenylalanine-tRNA synthetase which had a greatly reduced
ability to attach the analogue to tRNA (Fangman and Neidhardt, 1964).

2. Physiology of Mutants with Modified Aminoacyl-tRNA Synthetases

Mutants exist for several different aminoacyl-tRNA synthetases,
making it possible to examine the role of these enzymes in regulation
and cell physiology.

Temperature-conditional mutants for valine (Eidlic and Neidhardt,
1965; Yanif and Gros, 1966), alanine (Yanif and Gros, 1966) and
phenylalanine (Faiman et al., 1966) all show the same behaviour when
placed at the restrictive temperature: the accumulation of RNA ceases
as abruptly as the synthesis of protein. This would indicate that amino
acid-mediated regulation of RNA accumulation does not operate via
the free amino acid but through some derivative that requires a functional aminoacyl-tRNA synthetase. The same conclusion can be
drawn from the observation that p-fluoro-phenylalanine permits RNA
accumulation in a phenylalanine auxotroph which has normal phenylalanyl-tRNA synthetase but fails to do so in a mutant that cannot
activate the analogue (Fangman and Neidhardt, 1964).

This dependence of RNA synthesis on aminoacyl-tRNA synthetase
activity breaks down in cells having the "relaxed" allele of the RC-gene
(Stent and Brenner, 1961), where an overproduction of RNA occurs at
the restrictive temperature (Bock et al., 1966; Eidlic and Neidhardt,
1965). In a stringent cell both amino acids and their aminoacyl-tRNA
synthetases must be present for RNA to accumulate. It has been
suggested that the control mechanism works through the inhibition of
RNA synthesis by uncharged tRNA (Stent and Brenner, 1961;

Kurland and Maaloe, 1962; Tissières *et al.*, 1963). This possibility is made unlikely by the fact that the stringent phenylalanyl-tRNA synthetase mutant stops RNA synthesis at 40°, despite the absence of detectable uncharged tRNA (Neidhardt, 1966). The possibility that the concomitant synthesis of a particular protein is required for RNA synthesis has not yet been eliminated, largely because of the difficulty of establishing conditions giving no protein synthesis.

Gallant and Cashel (1967) have shown that the amino acid-dependent step is not the RNA polymerase reaction but rather the formation of UTP from UMP. Examination of the control of this reaction *in vitro* should help to determine the exact nature of the regulation of RNA synthesis.

When most temperature-sensitive aminoacyl-tRNA synthetase mutants are placed at their restricting temperatures, DNA synthesis continues after protein and RNA synthesis have halted. After increasing the total DNA by approximately 40%, synthesis ceases abruptly (Neidhardt and Faiman, 1966). This observation is consistent with the view that amino acids are required to make a special protein involved in the initiation of DNA synthesis and without which new rounds of DNA replication cannot be initiated (Lark, 1966).

3. Aminoacyl-tRNA Synthetases and Repression

Observations on several biosynthetic enzyme systems indicate that a functional aminoacyl-tRNA synthetase is required for the specific repression mechanism to operate. When a valyl-tRNA synthetase mutant is placed at a restrictive temperature, valine-controlled enzymes are derepressed (Eidlic and Neidhardt, 1965). The particular mutant used retains considerable ability to catalyse valyl adenylate formation *in vitro*, suggesting that the rate-limiting process at high temperature is the tRNA-attachment function of the enzyme. This then makes unlikely the possibility that the valyl–enzyme complex, or the valyl–adenylate–enzyme complex, acts as the repressor (or co-repressor).

There is very strong evidence showing that histidyl-tRNA synthetase is required to convert histidine into an active co-repressor. Schlesinger and Magasanik (1964) have shown that α-methyl histidine inhibits histidyl-tRNA synthetase in *Aerobacter aerogenes* and *E. coli*, leading to accumulation of histidine and derepression of the histidine-biosynthesizing enzymes. Amongst several classes of mutants that are derepressed for histidine biosynthesis in *Salmonella typhimurium*, one has a damaged histidyl-tRNA synthetase with an altered Km for histidine

(Roth and Ames, 1966). A second class, *his R*, has approximately 55% of the wild-type amount of histidine-accepting tRNA (Silbert *et al.*, 1966). It is not yet known whether the remaining activity is due to a second species of histidine-tRNA, or to a partial inactivation of the accepting ability of a single tRNA species. These findings suggest that histidyl-tRNA, rather than the free amino acid, is involved in the regulation of histidine biosynthesis. This could mean either that charged histidyl-tRNA acts as co-repressor of the histidine operon, or that the synthesis of a special protein is required for repression. Alternative models in which histidyl-tRNA is involved directly in control at the translational level have been described (Roth *et al.*, 1966).

In summary, it is evident that charged aminoacyl-tRNA, at least for valine and histidine, plays a role in two important regulatory functions; control of amino acid biosynthesis and control of RNA biosynthesis. The nature of the effector in each case remains to be elucidated and the possibility that it is an aminoacyl-tRNA species itself has not been eliminated.

REFERENCES

Allen, M. K. and Yanofsky, C. (1963). *Genetics, Princeton* **48**, 1065.
Ames, B. N. and Hartman, P. E. (1963). *Cold Spring Harb. Symp. quant. Biol.* **28**, 349.
Apirion, D. (1966). *J. molec. Biol.* **16**, 285.
Apirion, D. (1967). *J. molec. Biol.* **30**, 255.
Attardi, G. S., Naono, S., Rouviere, J., Jacob, F. and Gros, F. (1963). *Cold Spring Harb. Symp. quant. Biol.* **28**, 363.
Barnett, L., Brenner, S., Crick, F. H. C., Shulman, R. G. and Watts-Tobin, R. J. (1967). *Phil. Trans. R. Soc.* B**252**, 487.
Beckwith, J. R. (1963). *Biochim. biophys. Acta* **76**, 162.
Beckwith, J. R. and Signer, E. R. (1966). *J. molec. Biol.* **19**, 254.
Beckwith, J. R., Signer, E. R. and Epstein, W. (1966). *Cold Spring Harb. Symp. quant. Biol.* **31**, 393.
Benzer, S. and Champe, S. P. (1962). *Proc. natn. Acad. Sci. U.S.A.* **48**, 1114.
Berger, H. and Yanofsky, C. (1967). *Science, N.Y.* **156**, 394.
Berger, H., Brammar, W. J. and Yanofsky, C. (1968). *J. molec. Biol.* **34**, 219.
Bishop, J., Leahy, J. and Schweet, R. S. (1960). *Proc. natn. Acad. Sci. U.S.A.* **46**, 1030.
Böck, A., Faiman, L. E. and Neidhardt, F. C. (1966). *J. Bact.* **92**, 1076.
Bourgeois, S., Cohn, M. and Orgel, L. E. (1965). *J. molec. Biol.* **14**, 300.
Brammar, W. J., Berger, H. and Yanofsky, C. (1967). *Proc. natn. Acad. Sci. U.S.A.* **58**, 1499.
Bremer, H., Konrad, M. W., Gaines, K. and Stent, G. S. (1965). *J. molec. Biol.* **13**, 540.
Brenner, S. (1966). *Proc. R. Soc.* B**164**, 170.
Brenner, S. and Beckwith, J. (1965). *J. molec. Biol.* **13**, 629.
Brenner, S. and Stretton, A. O. W. (1965). *J. molec. Biol.* **13**, 944.

Brenner, S., Streisinger, G., Horne, R. W., Champe, S. P., Barnett, L., Benzer, S. and Rees, M. W. (1959). *J. molec. Biol.* **1**, 281.

Brenner, S., Barnett, L., Crick, F. H. C. and Orgel, A. (1961). *J. molec. Biol.* **3**, 121.

Brenner, S., Stretton, A. O. W. and Kaplan, S. (1965). *Nature, Lond.* **206**, 994.

Brenner, S., Kaplan, S. and Stretton, A. O. W. (1966). *J. molec. Biol.* **18**, 574.

Brenner, S., Barnett, L., Katz, E. R. and Crick, F. H. C. (1967). *Nature, Lond.* **213**, 449.

Bretscher, M. S. (1963). *J. molec. Biol.* **7**, 446.

Bretscher, M. S. and Jones, O. W. (1967). *In,* "Techniques in Protein Biosynthesis", P. N. Campbell and J. R. Sargent, eds.). Vol. I, p. 218, Academic Press, London.

Brody, S. and Yanofsky, C. (1963). *Proc. natn. Acad. Sci., U.S.A.* **50**, 9.

Brown, J. L., Brown, D. M. and Zabin, I. (1967). *Proc. natn. Acad. Sci. U.S.A.* **58**, 1139.

Campbell, A. (1962). *Adv. Genet.* **11**, 101.

Carbon, J., Berg, P. and Yanofsky, C. (1966). *Cold Spring Harb. Symp. quant. Biol.* **31**, 487.

Capecchi, M. R. (1967a). *Biochem. biophys. Res. Commun.* **28**, 773.

Capecchi, M. R. (1967b). *Proc. natn. Acad. Sci. U.S.A.* **58**, 1144.

Capecchi, M. R. and Gussin, G. N. (1965). *Science, N.Y.* **149**, 417.

Champe, S. P. and Benzer, S. (1962). *Proc. natn. Acad. Sci. U.S.A.* **48**, 432.

Chang, F. N., Sih, C. J. and Weisblum, B. (1966). *Proc. natn. Acad. Sci. U.S.A.* **55**, 431.

Clarke, P. H. and Brammar, W. J. (1964). *Nature, Lond.* **203**, 1153.

Cohn, M. and Horibata, K. (1959). *J. Bact.* **78**, 601.

Contesse, G., Naono, S. and Gros, F. (1966). *C. r. hebd. Séanc. Acad. Sci., Paris* **263**, 1007.

Cox, E. C., White, J. R. and Flaks, J. G. (1964). *Proc. natn. Acad. Sci. U.S.A.* **51**, 703.

Crick, F. H. C. (1963). *Prog. Nucleic Acid Res.* **1**, 164.

Crick, F. H. C. (1966). *J. molec. Biol.* **19**, 548.

Crick, F. H. C., Barnett, L., Brenner, S. and Watts-Tobin, R. J. (1961). *Nature, Lond.* **192**, 1227.

Cuzin, F. and Jacob, F. (1964). *C. r. hebd. Séanc. Acad. Sci., Paris* **258**, 1350.

Davies, J. E., Gilbert, W. and Gorini, L. (1964). *Proc. natn. Acad. Sci. U.S.A.* **51**, 883.

Davies, J. E., and Jacob, F. (1968). Quoted in Ippen *et al.* (1968).

Davies, J. E., Gorini, L. and Davis, B. D. (1965). *J. molec. Pharmacol.* **1**, 93.

Douglas, H. C. and Hawthorne, D. C. (1964). *Genetics, Princeton* **49**, 837.

Douglas, H. C. and Hawthorne, D. C. (1966). *Genetics, Princeton* **54**, 911.

Dove, W. F. (1966). *J. molec. Biol.* **19**, 187.

Drapeau, G. R., Brammar, W. J. and Yanofsky, C. (1968). *J. molec. Biol.* **35**, 357.

Eggertson, G. (1968). *Genet. Res.* **11**, 15.

von Ehrenstein, G. (1966). *Cold Spring Harb. Symp. quant. Biol.* **31**, 705.

Eidlic, L. and Neidhardt, F. C. (1965). *Proc. natn. Acad. Sci. U.S.A.* **53**, 539.

Engelhardt, D. L., Webster, R. E., Wilhelm, R. C. and Zinder, N. D. (1965). *Proc. natn. Acad. Sci. U.S.A.* **54**, 1791.

Engelhardt, D. L., Webster, R. E. and Zinder, N. D. (1967). *J. molec. Biol.* **29**, 45.

Englesberg, E. (1961). *J. Bacteriol.* **81**, 996.

Englesberg, E., Irr, J., Power, J. and Lee, N. (1965). *J. Bacteriol.* **90**, 946.

Epstein, R. H., Bolle, A., Steinberg, C. M., Kellenberger, E., Boy de la Tour, E., Chevalley, R., Edgar, R. S., Susman, M., Denhardt, G. and Lielavsis, A. (1963). *Cold Spring Harb. Symp. quant. Biol.* **28**, 375.

Faiman, L. E. (1966). Quoted in Neidhardt (1966).

Faiman, L. E., Böck, A. and Neidhardt, F. C. (1966). Quoted in Neidhardt (1966).

Fangman, W. L. and Neidhardt, F. C. (1964). *J. biol. Chem.* **239**, 1839.

Fink, G. R. and Martin, R. G. (1967). *J. molec. Biol.* **30**, 97.

Flaks, J. G., Cox, E. C. and White, J. R. (1962). *Biochem. biophys. Res. Commun.* **7**, 385.

Flaks, J. G., Leboy, P. S., Birge, E. A. and Kurland, C. G. (1966). *Cold Spring Harb. Symp. quant. Biol.* **31**, 623.

Fox, C. F., Beckwith, J. R., Epstein, W. and Signer, E. R. (1966). *J. molec. Biol.* **19**, 576.

Fowler, A. V. and Zabin, I. (1966). *Science, N.Y.* **154**, 1027.

Franklin, N. C. and Luria, S. E. (1961). *Virology* **15**, 299.

Freese, E. (1959a). *J. molec. Biol.* **1**, 87.

Freese, E. (1959b). *Proc. natn. Acad. Sci. U.S.A.* **45**, 622.

Freese, E., Bautz, E. and Freese, E. B. (1961). *Proc. natn. Acad. Sci. U.S.A.* **47**, 845.

Gallant, J. and Cashel, M. (1967). *J. molec. Biol.* **25**, 545.

Gallucci, E. and Garen, A. (1966). *J. molec. Biol.* **15**, 193.

Garen, A. and Siddiqi, O. (1962). *Proc. natn. Acad. Sci. U.S.A.* **47**, 1011.

Garen, A., Garen, S. and Wilhelm, R. C. (1965). *J. molec. Biol.* **14**, 167.

Gesteland, R. F., Salser, W. and Bolle, A. (1967). *Proc. natn. Acad. Sci. U.S.A.* **58**, 2036.

Gilbert, W. (1963). *J. molec. Biol.* **6**, 389.

Gilbert, W. and Muller-Hill, B. (1966). *Proc. natn. Acad. Sci. U.S.A.* **56**, 1891.

Gilbert, W. and Muller-Hill, B. (1967). *Proc. natn. Acad. Sci. U.S.A.* **58**, 2415.

Ghosh, H. P., Soll, D. and Khorana, H. G. (1967). *J. molec. Biol.* **25**, 275.

Goldberg, E. B. (1966). *Proc. natn. Acad. Sci. U.S.A.* **56**, 1457.

Goldstein, A., Kirschbaum, J. and Roman, A. (1965). *Proc. natn. Acad. Sci. U.S.A.* **54**, 1669.

Gorini, L. and Kataja, E. (1964). *Proc. natn. Acad. Sci. U.S.A.* **51**, 487.

Guest, J. R. and Yanofsky, C. (1965). *J. biol. Chem.* **240**, 679.

Guest, J. R. and Yanofsky, C. (1966). *Nature, Lond.* **210**, 799.

Gupta, N. K. and Khorana, H. G. (1966). *Proc. natn. Acad. Sci. U.S.A.* **56**, 772.

Gussin, G. N. (1966). *J. molec. Biol.* **21**, 435.

Helinski, D. R. and Yanofsky, C. (1962). *Proc. natn. Acad. Sci. U.S.A.* **48**, 173.

Helling, R. B. and Weinberg, R. (1963). *Genetics, Princeton* **48**, 1397.

Henning, U. and Yanofsky, C. (1962). *Proc. natn. Acad. Sci. U.S.A.* **48**, 183.

Henning, U., Dennert, G., Hertel, R. and Shipp, W. S. (1966). *Cold Spring Harb. Symp. quant. Biol.* **31**, 227.

Hogness, D. S., Doefler, W., Egan, J. B. and Black, L. W. (1966). *Cold Spring Harb. Symp. quant. Biol.* **31**, 129.

Horiuchi, K., Lodish, H. F. and Zinder, N. D. (1966). *Virology* **28**, 438.

Imamoto, F. and Yanofsky, C. (1967a). *J. molec. Biol.* **28**, 1.

Imamoto, F. and Yanofsky, C. (1967b). *J. molec. Biol.* **28**, 24.

Imamoto, F., Morikawa, N., Sato, K., Mishima, S. and Nishimura, T. (1965a). *J. molec. Biol.* **13**, 157.

Imamoto, F., Morikawa, N. and Sato, K. (1965b). *J. molec. Biol.* **13**, 169.

Imamoto, F., Ito, J. and Yanofsky, C. (1966). *Cold Spring Harb. Symp. quant. Biol.* **31**, 235.

Inouye, M., Akaboshi, E., Tsugita, A., Streisinger, G. and Okada, Y. (1967). *J. molec. Biol.* **30**, 39.

Ippen, K., Miller, J. H., Scaife, J. and Beckwith, J. R. (1968). *Nature, Lond.* **217**, 825.

Ito, J. and Crawford, I. P. (1965). *Genetics, Princeton* **52**, 1303.

Jacob, F. and Monod, J. (1961). *J. molec. Biol.* **3**, 318.

Jacob, F., Perrin, D., Sanchez, C. and Monod, J. (1960). *C. r. hebd. Séanc. Acad. Sci., Paris* **250**, 1727.

Jacob, F., Ullman, A. and Monod, J. (1964). *C. r. hebd. Séanc. Acad. Sci., Paris* **258**, 3125.

Jacob, F., Ullman, A. and Monod, J. (1965). *J. molec. Biol.* **13**, 704.

Jones, D. S., Nishimura, S. and Khorana, H. G. (1966). *J. molec. Biol.* **16**, 454.

Joyner, A., Isaacs, L. N. and Echols, H. (1966). *J. molec. Biol.* **19**, 174.

Kaiser, A. D. and Hogness, D. S. (1960). *J. molec. Biol.* **2**, 392.

Kaplan, S., Stretton, A. O. W. and Brenner, S. (1965). *J. molec. Biol.* **14**, 528.

Kiho, Y. and Rich, A. (1966). *Proc. natn. Acad. Sci. U.S.A.* **54**, 1751.

Koch, G. and Hershey, A. D. (1959). *J. molec. Biol.* **1**, 260.

Kurland, C. G. and Maaloe, O. (1962). *J. molec. Biol.* **4**, 193.

Lamfrom, H., MacLaughlin, C. S. and Sarabhai, A. S., (1966). *J. molec. Biol.* **22**, 355.

Landy, A., Abelson, J. N., Goodman, H. M. and Smith, J. D. (1967). *J. molec. Biol.* **29**, 457.

Lark, K. G. (1966). *Bact. Rev.* **30**, 3.

Lehmann, H. and Huntsman, R. G. (1966). "Man's Haemoglobin", North Holland Publishing Company, Amsterdam.

Lodish, H. F. and Zinder, N. D. (1966). *Science, N.Y.* **152**, 372.

Loomis, W. F. and Magasanik, B. (1964). *J. molec. Biol.* **8**, 417.

Loomis, W. F. and Magasanik, B. (1965). *Biochem. biophys. Res. Commun.* **20**, 230.

Loomis, W. F. and Magasanik, B. (1967). *J. molec. Biol.* **23**, 487.

MacDonald, R. E., Turnock, G. and Forchhammer, J. (1967). *Proc. natn. Acad. Sci. U.S.A.* **57**, 141.

Magasanik, B. (1961). *Cold Spring Harb. Symp. quant. Biol.* **26**, 249.

Maitra, U. and Hurwitz, J. (1965). *Proc. natn. Acad. Sci. U.S.A.* **54**, 815.

Malamy, M. H. (1966). *Cold Spring Harb. Symp. quant. Biol.* **31**, 189.

Mandelstam, J. (1961). *Biochem. J.* **79**, 489.

Marcker, K. and Sanger, F. (1964). *J. molec. Biol.* **8**, 835.

Margolin, P. (1965). *Science, N.Y.* **147**, 1456.

Margolin, P. and Bauerle, R. H. (1966). *Cold Spring Harb. Symp. quant. Biol.* **31**, 311.

Margolin, P. and Mukai, F. H. (1964). *Bact. Proc.* p. 87.

Martin, R. G. (1967). *J. molec. Biol.* **26**, 311.

Martin, R. G., Silbert, D. F., Smith, D. W. E. and Whitfield, H. J. (1966). *J. molec. Biol.* **21**, 357.

McFall, E. and Mandelstam, J. (1963a). *Nature, Lond.* **197**, 880.

McFall, E. and Mandelstam, J. (1963b). *Biochem. J.* **89**, 391.

Matsushiro, A., Sato, K., Ito, J., Kida, S. and Imamoto, F. (1965). *J. molec. Biol.* **11**, 54.

Morgan, A. R., Wells, R. D. and Khorana, H. G. (1967). *Proc. natn. Acad. Sci. U.S.A.* **56**, 1899.

Mosig, G. (1966). *Proc. natn. Acad. Sci. U.S.A.* **56**, 1177.

Mukai, F. H. and Margolin, P. (1963). *Proc. natn. Acad. Sci. U.S.A.* **50**, 140.

Muller-Hill, B. (1966). *J. molec. Biol.* **15**, 374.

Nathans, D., Notani, G., Schwartz, J. H. and Zinder, N. D. (1962). *Proc. natn. Acad. Sci. U.S.A.* **48**, 1424.

Neidhardt, F. C. (1966). *Bact. Rev.* **30**, 701.

Neidhardt, F. C. and Faiman, L. E. (1966). Quoted in Neidhardt (1966).

Newton, W. A., Beckwith, J. R., Zipser, D. and Brenner, S. (1965). *J. molec. Biol.* **14**, 290.

Nirenberg, M. and Leder, P. (1964). *Science, N.Y.* **145**, 1399.

Nirenberg, M. W. and Matthaei, J. H. (1961). *Proc. natn. Acad. Sci. U.S.A.* **47**, 1588.

Nishimura, S., Jones, D. S. and Khorana, H. G. (1965a). *J. molec. Biol.* **13**, 302.

Nishimura, S., Jones, D. S., Ohtsuka, E., Hayatsu, H., Jacob, T. M. and Khorana, H. G. (1965b). *J. molec. Biol.* **13**, 283.

Nomura, M. and Lowry, C. V. (1967). *Proc. natn. Acad. Sci. U.S.A.* **58**, 946.

Notani, G. W., Engelhardt, D. L., Konigsberg, W. and Zinder, N. D. (1965). *J. molec. Biol.* **12**, 439.

Novotny, C. and Englesberg, E. (1964). *Proc. Int. Congr. Biochem.* 6th, New York, **3**, 46.

Okada, Y., Terzaghi, E., Streisinger, G., Emrich, J., Inouye, M. and Tsugita, A. (1966). *Proc. natn. Acad. Sci. U.S.A.* **56**, 1692.

Palmer, J. and Moses, V. (1967). *Biochem. J.* **103**, 358.

Palmer, J. and Moses, V. (1968). *Biochem. J.* **106**, 339.

Power, J. (1967). *Genetics, Princeton* **55**, 557.

Ptashne, M. (1967a). *Proc. natn. Acad. Sci. U.S.A.* **57**, 306.

Ptashne, M. (1967b). *Nature, Lond.* **214**, 232.

Rifkin, D. B., Hirsh, D. I., Rifkin, M. R. and Konigsberg, W. (1966). *Cold Spring Harb. Symp. quant. Biol.* **31**, 715.

Roth, J. R. and Ames, B. N. (1966). *J. molec. Biol.* **22**, 325.

Roth, J. R., Anton, D. N. and Hartman, P. E. (1966). *J. molec. Biol.* **22**, 305.

Salas, M., Smith, M. A., Stanley, W. M., Wahba, A. J. and Ochoa, S. (1965). *J. biol. Chem.* **240**, 3988.

Sambrook, J. F., Fan, D. P. and Brenner, S. (1967). *Nature, Lond.* **214**, 452.

Sarabhai, A. S. and Brenner, S. (1967). *J. molec. Biol.* **27**, 145.

Sarabhai, A. S., Stretton, A. O. W., Brenner, S. and Bolle, A. (1964). *Nature, Lond.* **201**, 13.

Scaife, J. and Beckwith, J. R. (1966). *Cold Spring Harb. Symp. quant. Biol.* **31**, 403.

Schlesinger, S. and Magasanik, B. (1964). *J. molec. Biol.* **9**, 670.

Schlessinger, D., Mangiarotti, G. and Apirion, D. (1967). *Proc. natn. Acad. Sci. U.S.A.* **58**, 1782.

Sheppard, D. and Englesberg, E. (1966). *Cold Spring Harb. Symp. quant. Biol.* **31**, 345.

Sheppard, D. and Englesberg, E. (1967). *J. molec. Biol.* **25**, 443.

Signer, E. R., Beckwith, J. R. and Brenner, S. (1965). *J. molec. Biol.* **14**, 153.

Silbert, D. F., Fink, G. R. and Ames, B. N. (1966). *J. molec. Biol.* **22**, 335.

Smith, J. D., Abelson, J. N., Clark, B. F. C., Goodman, H. M. and Brenner, S. (1966). *Cold Spring Harb. Symp. quant. Biol.* **31**, 479.

Smith, O. H. and Yanofsky, C. (1963). *In*, "Methods in Enzymology" (S.P. Colowick and N. O. Kaplan, eds.). Vol. 5, p. 794, Academic Press, New York and London.

Speyer, J. F., Lengyel, P., Basilio, C. and Ochoa, S. (1962). *Proc. natn. Acad. Sci. U.S.A.* **48**, 63.

Staehelin, T. and Meselson, M. (1966). *J. molec. Biol.* **19**, 207.

Stahl, F. W., Murray, N. E., Nakata, A. and Crasemann, J. M. (1966). *Genetics, Princeton* **54**, 223.

Stent, G. S. (1964). *Science, N.Y.* **144**, 816.

Stent, G. S. and Brenner, S. (1961). *Proc. natn. Acad. Sci. U.S.A.* **47**, 2005.

Streisinger, G., Okada, Y., Emrich, J., Newton, J., Tsugita, A., Terzaghi, E. and Inouye, M. (1966). *Cold Spring Harb. Symp. quant. Biol.* **31**, 77.

Streisinger, G., Emrich, J., Okada, Y., Tsugita, A. and Inouye, M. (1968). *J. molec. Biol.* **31**, 607.

Stretton, A. O. W. and Brenner, S. (1965). *J. molec. Biol.* **12**, 456.

Stretton, A. O. W., Kaplan, S. and Brenner, S. (1966). *Cold Spring Harb. Symp. quant. Biol.* **31**, 173.

Taubman, S. B., Jones, N. R., Young, F. E. and Corcoran, J. W. (1966). *Biochim. biophys. Acta* **123**, 438.

Taylor, A. L. and Trotter, C. D. (1967). *Bact. Rev.* **31**, 332.

Terzaghi, E., Okada, Y., Streisinger, G., Emrich, J., Inouye, M. and Tsugita, A. (1966). *Proc. natn. Acad. Sci. U.S.A.* **56**, 500.

Thomas, R. (1966). *J. molec. Biol.* **22**, 79.

Thach, R. E., Cecere, M. A., Sundarajan, T. A. and Doty, P. (1965). *Proc. natn. Acad. Sci. U.S.A.* **54**, 1167.

Tissières, A., Bourgeois, S. and Gros, F. (1963). *J. molec. Biol.* **7**, 100.

Traub, P., Hosokawa, K. and Nomura, M. (1966). *J. molec. Biol.* **19**, 211.

Tsugita, A., Inouye, M., Terzaghi, E. A., Emrich, J., Okada, Y. and Streisinger, G. (1967). Abstracts of the *7th Int. Congr. Biochem.* **1**, 19.

Van Vunakis, H., Baker, W. H. and Brown, R. K. (1958). *Virology* **5**, 327.

Watson, J. D. and Crick, F. H. C. (1953). *Nature, Lond.* **171**, 737.

Weigert, M. G. and Garen, A. (1965). *Nature, Lond.* **206**, 992.

Weigert, M. G., Lanka, E. and Garen, A. (1965). *J. molec. Biol.* **14**, 522.

Weigert, M. G., Gallucci, E., Lanka, E. and Garen, A. (1966). *Cold Spring Harb. Symp. quant. Biol.* **31**, 145.

Weigert, M. G., Lanka, E. and Garen, A. (1967). *J. molec. Biol.* **23**, 401.

Whitfield, H. J., Martin, R. G. and Ames, B. N. (1966). *J. molec. Biol.* **21**, 335.

Wilhelm, R. C. (1966) in discussion of Carbon, J., Berg, P. and Yanofsky, C. (1966). *Cold Spring Harb. Symp. quant. Biol.* **31**, 487.

Willson, C., Perrin, D., Cohn, M., Jacob, F. and Monod, J. (1964). *J. molec. Biol.* **8**, 582.

Wittmann, H. G. and Wittmann-Liebold, B. (1966). *Cold Spring Harb. Symp. quant. Biol.* **31**, 163.

Woese, C. R. (1967). *In*, "Progress in Nucleic Acid Research" (J. N. Davidson and W. E. Cohn, eds.). Vol. 7, p. 107, Academic Press, New York and London.

Yanif, M. and Gros, F. (1966). Quoted in Neidhardt (1966).

Yanofsky, C. and Ito, J. (1966). *J. molec. Biol.* **21**, 313.

Yanofsky, C. and Ito, J. (1967). *J. molec. Biol.* **24**, 143.

Yanofsky, C., Carlton, B. C., Guest, J. R., Helinski, D. R. and Henning, U. (1964). *Proc. natn. Acad. Sci. U.S.A.* **51**, 266.

Yanofsky, C., Cox, E. C. and Horn, V. (1966a). *Proc. natn. Acad. Sci. U.S.A.* **55**, 274.

Yanofsky, C., Ito, J. and Horn, V. (1966b). *Cold Spring Harb. Symp. quant. Biol.* **31**, 151.

Yanofsky, C., Drapeau, G. R., Guest, J. R. and Carlton, B. C. (1967). *Proc. natn. Acad. Sci. U.S.A.* **57**, 296.

Zinder, N. D., Engelhardt, D. L. and Webster, R. E. (1966). *Cold Spring Harb. Symp. quant. Biol.* **31**, 251.

Zipser, D. (1967). *J. molec. Biol.* **29**, 441.

Zwaig, N. and Lin, E. C. C. (1965). *Biochem. biophys. Res. Commun.* **22**, 414.

ACKNOWLEDGEMENTS

I would like to thank Professor M. R. Pollock for his helpful suggestions and critical reading of the manuscript.

The author is a member of the Medical Research Council Group on Bacterial Enzyme Variation at the University of Edinburgh, Scotland.

CHAPTER 2

Protein Biosynthesis in Plant Systems

JORGE E. ALLENDE

*Departamento de Química, Facultad de Ciencias,
Universidad de Chile, Santiago, Chile*

I. INTRODUCTION

During the last 15 years, biochemists and molecular biologists have mounted a massive and sustained effort to understand the cellular process of protein biosynthesis. The spectacular advances made in this field have been made by investigators working with bacterial and animal systems. Work on protein synthesis in plants has lagged considerably and the number of scientists involved has been much smaller. Nevertheless, there is solid evidence to demonstrate that the overall process of protein biosynthesis in plants is similar to that found in other species.

Now that the emphasis of the research in this field is shifting towards the study of the regulation of protein synthesis in higher organisms, one would imagine that the study of plant systems would gain popularity. Plants present some unique phenomena that lend themselves to experimental investigation at the molecular level, such as the totipotentiality of some of their cells, the trigger mechanisms that result in bursts of fantastic metabolic activity after prolonged periods of dormancy, the effects of hormonal substances on the process of germination, the ripening of fruits and the shedding of the leaves, and the action of light upon the appearance of the chloroplasts as distinct and functional cellular organelles. In addition, in a more pragmatic sense, the study of protein synthesis in plants should be motivated by the idea that vegetable proteins of high nutritional value may represent the only solution to the problem of feeding an over-populated world.

This chapter includes only a description of the methods of preparing the plant components that intervene in *in vitro* protein synthesis. It does not include methods of isolation and characterization of plant messenger RNA because of the scarcity of published material on this subject. Recent and thorough reviews on the subject of protein synthesis in plants have been written by Mans (1967) and by Holley (1965).

II. RIBOSOMES

Ribosomes (microsomal particles, Palade granules) are particles composed of approximately equal proportions of protein and ribonucleic acid and are found in the cytoplasm as well as in some cellular organelles. Following their discovery with the electron microscope (Robinson and Brown, 1953; Palade, 1955), Littlefield *et al.* (1955) established the involvement of these particles in the process of protein synthesis in mammalian liver. Webster (1957) soon after described a similar function in plant cells.

A. Isolation of Plant Ribosomes

During the last few years ribosomes have been isolated from many different plant sources. Most methods for this isolation, however, are very similar. The following procedure has been used in our laboratory for the preparation of ribosomes from viable wheat embryos isolated by the procedure of Johnston and Stern (1957). Wheat embryos (20 g) which had been kept frozen at $-20°$ are ground in a cold mortar with washed sea sand in the presence of a buffer containing 0·5M sucrose, 50 mM Tris-HCl, pH 7·5, 10 mM $MgCl_2$, 25 mM KCl and 5 mM 2-mercaptoethanol. All manipulations should be performed at 0–4°. At the start of the grinding, only 5 ml of this buffer are mixed with the tissue and as the grinding proceeds more buffer is added gradually until 40 ml of the solution are present in the homogenate. Grinding should continue until a homogeneous paste has been obtained. This paste is passed through four layers of gauze, forcing through the liquid retained in the gauze in order to reduce the losses.

The filtrate is subsequently centrifuged at 30,000g for 20 min in a refrigerated centrifuge. The supernatant fluid is again passed through two layers of gauze to retain an oily scum that floats on the liquid. The supernatant liquid (fairly clear) is then centrifuged at 105,000g for 150 min in the ultracentrifuge. The supernatant obtained is stored frozen as a source of aminoacyl-tRNA synthetases and transfer enzymes. (Immediately before using this supernatant it can be passed through a column of Sephadex G-25 equilibrated with 10 mM Tris, pH 7·5, and 1 mM 2-mercaptoethanol to eliminate endogenous amino acids and nucleotides. The protein emerges turbid from the Sephadex and the synthetase enzymes present seem to be much less stable than in the original supernatant.) The ribosomal pellet is carefully resuspended in 5 ml of the same buffer used for grinding. A homogeneous suspension is obtained by using a glass homogenizer. The suspension is centrifuged again at 30,000g for 20 min to eliminate aggregates. This ribosomal preparation can be stored as such, freezing in small aliquots. Freezing in liquid nitrogen seems to be best for preserving the activity of the particles. If "washed" ribosomes are desired to study transfer enzymes, the ribosomes are diluted with the same grinding buffer, centrifuged again for 150 min at 105,000g and the pellet resuspended in minimal buffer to obtain as concentrated a ribosomal suspension as possible (higher than 10 mg of ribosomal RNA/ml). These ribosomes are completely dependent on transfer enzymes for activity and can be stored for several weeks at $-20°$. In tissues where the ribosomal particles are bound to the membranes of the endoplasmic reticulum (microsomes), the first washing of the ribosomes must be performed in the presence of 0·2% deoxycholate (Mans and Novelli, 1964). This detergent frees the particles from the membranes without destroying their activity but deoxycholate must be eliminated from the ribosomes by two subsequent re-sedimentations in deoxycholate-free buffer.

B. Properties of Plant Ribosomes

Plant ribosomes contain from 40–50% RNA and from 50–60% protein. They are similar to the particles from other sources in that they are composed of subunits which can be dissociated and reassociated by changing the nature of the medium (T'so et al., 1956; T'so, 1962; Bayley, 1964). In the presence of 5×10^{-3} M Mg^{++} the particles have a sedimentation coefficient of 80s and under the electron microscope appear as oblate spheroids with a diameter of 250 Å. Removal of Mg^{++} by dialysis causes their dissociation into a 60s subunit and a 40s subunit. These two subunits reassociate to the original 80s particle by restoring the magnesium concentration to 10^{-3} M. Further removal of magnesium, by addition of the chelating agent EDTA at 10^{-2} M, causes dissociation of the 60s particle into even smaller subunits with a sedimentation coefficient of 26s.

Deproteinization of ribosomal particles by the phenol method results in the isolation of ribosomal RNA (Wallace et al., 1961). Analysis of the RNA obtained by this method showed that there are three components with different sedimentation coefficients: 28s, 18s and approximately 4s. The 28s RNA is present in the 60s ribosomal subunit, the 18s RNA originates in the 40s subunit and the 4s material may resemble the 5s material found in Escherichia coli (Rosset et al., 1964). The ribosomes from peas have also been studied in regard to their proteins. The proteins present in the particles are basic in nature, seem to be a mixture of different polypeptides of molecular weights between 12,000 and 25,000 and can be separated into several components by gel electrophoresis (Setterfield et al., 1960; Lyttleton, 1968).

The extent of attachment of ribosomes to the membranes of the endoplasmic reticulum to form microsomes varies with the plant tissue and also with the physiological state of the cell. Tobacco pith cells in culture (Flamm and Birnstiel, 1964) show a correlation between their rapid rate of growth and the high amount of ribosomes that are membrane bound. Senescent cells contain a higher proportion of free ribosomes. In contrast some merestimatic cells which are rapidly dividing have mostly free ribosomes, which later are bound to the membrane as the cells mature (Whaley et al., 1960).

C. Polysomes

In the translation process of protein synthesis several ribosomal particles can attach to a molecule of messenger RNA increasing the efficiency of the overall reading process. The resulting complexes of several ribosomes bound to a molecule of messenger RNA have been called polysomes (Warner et al., 1962).

Since only the ribosomes that are actually attached to a messenger RNA are actively participating in protein synthesis, the amount of polysomes present in a tissue can be considered an indication of the protein synthesizing activity of the cells in that tissue.

Polysomes can be isolated and identified by sucrose density gradient centrifugation. The ribosomes which aggregate to form polysomes with two, three, four, five or more particles sediment faster than single 80s ribosomes (monosomes).

The main difficulty with studies of this type is the extreme sensitivity of polysomes to nucleases which cleave the messenger RNA and destroy the aggregates. Several reports of the existence of polyribosomes in plant tissues have been published (Clark et al., 1964; Dure and Waters, 1965). Marcus and Feeley (1965) found a clear correlation between the rapid increase in protein synthesis following water imbibition by dry peanut cotyledons and wheat embryos and the increased levels of polysomes present in the extracts of these tissues (Fig. 1). Destruction of mRNA by nucleases did not seem to occur in these extracts and no special precautions were required. However, other plant tissues such as barley embryos have appreciable nuclease activity so that poly-ribosomes are destroyed in the extraction procedure. A recent publication of Watts and Mathias (1967) showed that polyribosomes from this tissue can be preserved for analysis if the extraction is made in the presence of bentonite that has been graded as to particle size, and freed of metals by EDTA.

D. Chloroplast Ribosomes

Lyttleton (1962) found that isolated spinach leaf chloroplasts contain ribosomes with a sedimentation coefficient of 66s as compared to the 83s of the cytoplasmic particles. These chloroplast ribosomes yield two subunits of 47s and 33s upon removal of magnesium. In the case of *Euglena gracilis*, the ribosomes isolated from the chloroplasts are much smaller and their sedimentation coefficient is 44s (Brawerman, 1963). Chloroplast ribosomes have been shown to be active in *in vitro* systems of protein synthesis (Boardman et al., 1965; Eisenstadt and Brawerman, 1964; Boardman et al., 1966), and respond to plant aminoacyl transfer enzymes (Allende and Bravo, 1966). The isolation of ribosomal particles from the organelles has been described by Eisenstadt and Brawerman (1964), starting from purified *Euglena gracilis* chloroplasts. Chloroplasts, which had been kept in a buffer containing 10% sucrose, 10 mM Tris-HCl, pH 7·6, 4 mM $MgCl_2$ and 1 mM 2-mercaptoethanol, were transferred to a buffer containing 10 mM Tris

60   JORGE E. ALLENDE

buffer, pH 7·6, 10 mM MgCl$_2$ and 0·5% sodium deoxycholate to produce lysis of the organelles. The lysate was centrifuged at 23,000g for
30 min and ribosomes in the supernatant were pelleted by centrifuging
at 105,000g for 2 h. The ribosomal pellets were resuspended in buffer

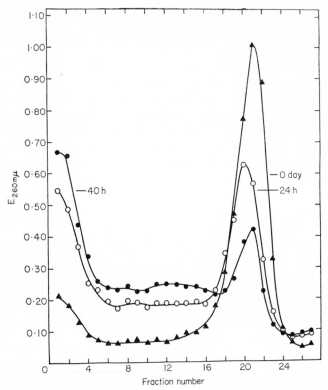

FIG. 1. Effect of water imbibition on the polysome content of peanut cotyledons.
The sedimentation patterns of the ribosomal fractions of peanut cotyledons at 0 h
after imbibition (▲—▲, 0·34 mg of ribosomal RNA), 24 h after imbibition
(○—○, 0·48 mg of ribosomal RNA) and 40 h after imbibition (●—●, 0·54 mg of
ribosomal RNA) were observed in a 5 to 20% sucrose density gradient. Redrawn
with permission from Marcus and Feeley (1965).

containing 10 mM Tris-HCl, pH 7·6, 2 mM MgCl$_2$, and centrifuged at
3000g for 10 min to remove contaminating green material. The
ribosomes obtained by this method were active in *in vitro* protein
synthesis.

III. AMINOACYL-tRNA SYNTHETASES

The aminoacyl-tRNA synthetases (amino acid activating enzymes or
aminoacyl-tRNA ligases) catalyse the reaction which is considered to be

the first step in protein biosynthesis. The reaction mechanism is known to require two steps to form aminoacyl-tRNA:

(1)   Amino acid + ATP + Enzyme
       ⇌ Aminoacyl–AMP–Enzyme + PPi

(2)   Aminoacyl–AMP–Enzyme + tRNA
       ⇌ Aminoacyl–tRNA + AMP + Enzyme

Three main methods for measuring the activities of these enzymes have been described. One measures the formation of aminoacyl hydroxamates in the presence of ATP, another determines the exchange of ^{32}P labelled pyrophosphate with ATP in the presence of the amino acid, and the third method measures the overall reaction, i.e. the formation of aminoacyl-tRNA in the presence of amino acid, ATP and tRNA.

A. Assays

1. The Hydroxamate Assay

The hydroxamate assay for aminoacyl-tRNA synthetases is described in detail by Stulberg and Novelli (1962). The reaction mixture contains: enzyme, 200 μmoles of Tris-HCl buffer, pH 7·8, 10 μmoles of ATP, 10 μmoles of the amino acid and 1000 μmoles of NH_2OH, salt-free as described by Beinert et al. (1953), in a volume of 1 ml. After incubating for 30 min at 37° the reaction is stopped by the addition of 2·3 ml of a mixture containing 0·37M $FeCl_3$, 0·31M trichloroacetic acid and 0·65M HCl. The precipitated protein is removed by centrifuging and the optical density of the supernatant fluid is measured at 540 mμ. Commercially available amino acid hydroxamates can be used as standards for the determination.

It is noteworthy that the presence of endogenous amino acids and other chromogenic substances in the crude enzyme preparations result in high blank values. Dialysis, gel filtration, or the use of other methods that eliminate amino acids and pigments from the enzyme preparation is advisable.

2. The Pyrophosphate Exchange Assay

The exchange of radioactive pyrophosphate with ATP was used as an assay for amino acid activation by DeMoss and Novelli (1956). Attwood and Cocking (1965) have applied it to the assay of alanyl-tRNA synthetase from tomato roots as follows. A mixture of 1 ml containing 10 μmoles of ATP (dipotassium salt), 9 μmoles of [^{32}P]-pyrophosphate (specific activity of 70,000 cpm/μmole), 100 μmoles of

L-alanine and enzyme preparation is incubated at 37° for 30 min. The reaction is stopped by the addition of 1 ml of 10% trichloroacetic acid and the precipitated protein removed by centrifuging. The ATP and the pyrophosphate in the supernatant are separated by addition of 4 ml of 0·1M sodium acetate and 100 mg of acid-washed charcoal (Norit-A). After mixing thoroughly, the Norit-A containing the ATP is sedimented by centrifuging. The charcoal is then washed three times with 0·05M sodium acetate buffer, pH 4·5, and once with distilled water. The ATP adsorbed onto the Norit-A is hydrolysed to AMP by adding 4·0 ml of 1·0N HCl and heating in a boiling water bath for 15 min. The mixture is cooled and centrifuged and radioactivity determined in aliquots of the supernatant. The results should be calculated using control determination carried out in the absence of the added amino acid.

Again in this assay the presence of endogenous amino acids in the enzyme preparations results in high blanks. The presence of ATPases is also obviously a complication. This latter problem has been diminished by addition of 50 μmoles of KF per ml of reaction mixture. These levels of KF inhibit most ATPases and do not affect amino acid activation. In some cases the pyrophosphate exchange reaction is greatly influenced by the presence of tRNA (Ravel et al., 1965; Lazzarini and Mehler, 1966).

3. The Formation of Aminoacyl-tRNA

In our laboratory we have used the system described below to measure the activities of the plant aminoacyl-tRNA synthetases specific for glycine, threonine, glutamic acid, aspartic acid, phenylalanine, serine, leucine, valine and methionine. The incubation medium of 1·0 ml contains 80 μmoles of Tris-HCl buffer, pH 7·5, 7·5 μmoles of MgCl$_2$, 4·5 μmoles of ATP, 3 μmoles of reduced glutathione, 50 mμmoles of each of the 19 other amino acids, 10 mμmoles of the amino acid to be tested labelled with ^{14}C at a specific activity of 25 μc/μmole, 0·1 mg of purified plant tRNA and enzyme. After incubating for 15 min at 37° the reaction is stopped by the addition of 4 ml of cold 5% trichloroacetic acid containing 0·001M of the non-radioactive amino acid that is being assayed. After keeping the suspension 10 min in ice, it is filtered on nitrocellulose filters (Millipore Corporation, 0·45 μ pore diameter) and the filter is washed three times with 4 ml of the same cold trichloroacetic acid solution. The dry filter may be counted on planchets in a gas flow counter or with a scintillation mixture in a scintillation spectrometer.

Sometimes this assay is complicated by the presence of high concentrations of ribonuclease in the enzyme preparations. In those cases, a partial purification of the synthetases by chromatography on a DEAE-cellulose column as described by Muench and Berg (1966) can be useful. In each case, of course, the chromatographic properties of the synthetases may vary. Using a 105,000g supernatant from wheat embryo extract (described in the preparation of ribosomes), we have not encountered ribonuclease in amounts sufficient to hinder the assay, and the only purification carried out has been to pass the supernatant fraction through a G-25 Sephadex column to eliminate endogenous

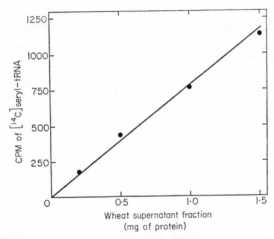

FIG. 2. Effect of enzyme concentration on the formation of [^{14}C]seryl-tRNA. The formation of [^{14}C]seryl-tRNA was assayed as described in the text in the presence of different amounts of wheat supernatant fraction that had been passed through Sephadex G-50.

amino acids. The graph in Fig. 2 shows the effect of increasing amounts of enzyme on the formation of [^{14}C]seryl-tRNA.

B. Properties of Plant Aminoacyl-tRNA Synthetases

Soon after the discovery of the aminoacyl-tRNA synthetases in mammalian (Hoagland, 1955) and bacterial systems (Berg, 1956), Webster (1957) presented evidence for the presence of the enzymes in pea microsomes and Clark (1958) detected their activity in spinach (*Spinacia oleracea*) leaves. Clark's work demonstrated by the pyrophosphate exchange assay that acetone powders of spinach leaves, *Avena* coleoptiles, asparagus tips, winter rye grass and tobacco leaves had the capacity to activate most amino acids. He also concluded from his experiments that the aminoacyl-tRNA synthetases were present in the cell in the soluble or 105,000*g* supernatant fraction. Moustafa and his collaborators (Moustafa and Proctor, 1962; Moustafa and Lyttleton, 1963) have obtained preparations that have amino acid activating activity from root nodules of *Galega officianalis* and from wheat germ. He also has achieved a considerable purification (300-fold) of the valine activating enzyme from wheat germ (Moustafa, 1963) and studied its properties. Attwood and Cocking (1965) have likewise obtained a highly purified preparation (3000-fold) of alanyl-

TABLE 1

Species specificity of aminoacyl-tRNA synthetases

A　Seryl-tRNA synthetases

Enzyme source	tRNA source	[^{14}C]aminoacyl-tRNA cpm
Wheat embryo	Wheat embryo	848
Wheat embryo	E. coli	257
E. coli	Wheat embryo	0
E. coli	E. coli	363
Yeast	Wheat embryo	665
Yeast	Yeast	579

B　Methionyl-tRNA synthetases

E. coli	Wheat embryo	4,905
E. coli	E. coli	10,791
Wheat embryo	Wheat embryo	7,812
Wheat embryo	E. coli	8,801

The formation of [^{14}C]aminoacyl-tRNAs was assayed by the method described in the text. In all cases 100 μg of purified tRNA were used and an amount of enzyme that had been found to be saturating in the homologous system. [^{14}C]Serine (specific activity 25 μc/μmole) and [^{14}C]methionine (specific activity 225 μc/μmole) were used.

tRNA synthetase from tomato roots. These authors found an unusually high Michaelis constant for alanine by both the hydroxamate and the pyrophosphate assays ($7 \cdot 1 \times 10^{-2}$ M and $2 \cdot 8 \times 10^{-2}$ M, respectively). The usual Km for the amino acid of these enzymes is of the order of 10^{-6} M (Schweet and Allen, 1958). The high Michaelis constant of some enzymes might explain why they appeared to be absent in some plant tissues (Davis and Novelli, 1958) when assayed under the usual conditions employing a concentration of 10^{-4} M for the amino acid.

The phenomenon of species specificity between aminoacyl-tRNA synthetases and tRNA preparations from different organisms is well documented (Doctor and Mudd, 1963). Little work has been done, however, to compare the compatibility of plant aminoacyl-tRNA synthetases and tRNAs with the components from other sources. Jacobson *et al.* (1964) did include tRNA and aminoacyl-tRNA synthetase preparations from maize in their specifity studies. From our work

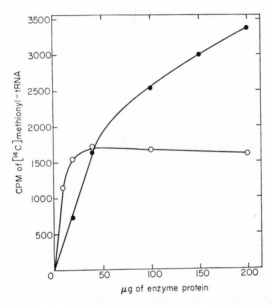

FIG. 3. Acylation of wheat tRNA with methionine by the methionyl-tRNA synthetases from wheat and from *E. coli*. The assay for the formation of [14C]-methionyl-tRNA was carried out as described in the text, except that only 50 μg of wheat tRNA were used and the [14C]methionine had a specific activity of 189 μc/μmole. In one experiment (●—●) the enzyme was a supernatant fraction from a wheat embryo extract, while in the other experiment (○—○) the enzyme was a tRNA-free preparation from *E. coli*.

with wheat embryo enzymes and tRNA we have found that, in some instances, the enzymes from a heterologous source charge the plant tRNA fully as is the case with the yeast seryl-tRNA synthetase (Table 1). In the same experiments it can be seen that the equivalent enzyme from wheat embryo charges *E. coli* tRNA partially, while the *E. coli* synthetase is completely unable to act on the wheat tRNA. In the case of the methionyl-tRNA synthetase, the *E. coli* enzyme charges only 50% of the wheat tRNA (Fig. 3). The explanation of this result is that the *E. coli* synthetase recognizes only one of the two species of methionyl-tRNA present in wheat embryo (Fig. 4).

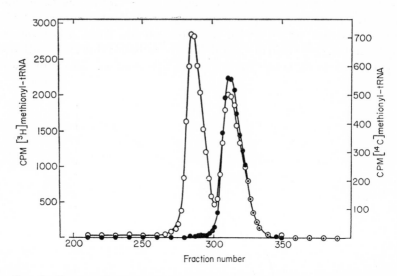

Fig. 4. Reverse-phase chromatography of wheat methionyl-tRNA acylated with wheat and *E. coli* methionyl-tRNA synthetases. Fifty mg of [³H]methionyl-tRNA were prepared with [³H]methionine (specific activity 1·4 μc/μmole) and wheat supernatant fraction. Fifty mg of [¹⁴C]methionyl-tRNA were likewise prepared with [¹⁴C]methionine (specific activity 224 μc/μmole) and with tRNA-free *E. coli* methionyl-tRNA synthetase. The two labelled compounds were treated with phenol and isolated as described in the text, mixed and applied to a reverse-phase chromatographic column (2·40 m × 2·5 cm) prepared according to Kelmers *et al.* (1965). The column was eluted with a linear salt gradient from 0·4M to 0·75M NaCl in a buffer solution containing 0·01M sodium acetate, pH 4·5, 0·01M magnesium acetate and 0·001M EDTA. Each gradient chamber contained 5 l and the flow was controlled to 6 ml/min. Fractions of 16 ml were collected and 300 μl aliquots were counted for ³H (○—○) and ¹⁴C (●—●) in the liquid scintillation mixture described by Bray (1960).

IV. TRANSFER RIBONUCLEIC ACID

Transfer RNAs (soluble RNAs) are the "bilingual" adaptor molecules that, through the action of the aminoacyl-tRNA synthetases, accept specific amino acids and transfer them to the nascent polypeptides as directed by the messenger RNA on the ribosomes. These interesting polynucleotides which have a molecular weight of approximately 25,000 were discovered first in a mammalian system (Hoagland *et al.*, 1958) and soon after in a bacterial extract (Berg and Ofengand, 1958). The formation of aminoacyl-tRNAs in plant systems was first reported by Webster (1959a) and by Ichimura *et al.* (1960). The purification of plant tRNAs has been achieved by methods adapted from those employed in bacterial and mammalian systems (Brunngraber, 1962; Monier et al., 1960).

A. Method for Isolation of tRNA from Wheat Germ

The following method to obtain tRNA preparations from commercial wheat germ has been used in our laboratory. Wheat germ in 200 g batches (not toasted, obtained from the International Milling Company, Minneapolis, Minnesota, or prepared as described by Johnston and Stern, 1957) is suspended in 500 ml of 0·01M Tris-HCl buffer, pH 7·5, and 0·1M NaCl, and homogenized with 3 volumes of water-saturated phenol in a stainless steel Waring blendor for 5 min. The two phases are separated by centrifuging at 15,000g for 20 min in a refrigerated centrifuge. The water phase is carefully freed of an oily top layer and is re-extracted with an equal volume of fresh, water-saturated phenol. The water layer obtained after a similar centrifugation is made 2% with respect to potassium acetate, pH 5·0, and nucleic acid precipitated with $2\frac{1}{2}$ volumes of 95% ethanol at $-20°$. The solution is left overnight at $-20°$ and the precipitate is collected by centrifuging (10,000g for 15 min at $-20°$).

The precipitate is then suspended in a buffer containing 0·01M Tris, pH 7·2, and 0·001M $MgCl_2$ using a volume equal to the initial weight of wheat. Crystalline pancreatic DNase (Worthington Biochemicals) is added to a final concentration of 2 $\mu g/ml$ of suspension. After 15 min at 37°, solid sodium chloride is added to a concentration of 2M. The suspension is allowed to stir for 2 or 3 h in the cold and is centrifuged at 0° for 20 min at 10,000g.

The precipitate containing the high molecular weight RNAs is discarded and the remaining supernatant is precipitated with potassium acetate and ethanol as before. The precipitate is then dissolved in cold 0·1M Tris, pH 7·5, and dialysed against the same buffer.

The transfer RNA is further purified by chromatography on a DEAE-cellulose column. A column of 15×4 cm can be used for a preparation containing 20,000A_{260} units of nucleic acid. The column is equilibrated with 0·1M Tris-HCl, pH 7·5, and, after loading with the tRNA preparation, it is extensively washed with the same buffer (1·5 l). The tRNA is eluted with a buffer containing 0·1M Tris, pH 7·5, 1·0M NaCl and 0·005M EDTA, and the eluent is collected until the absorbance at 260 mμ falls to less than 2A_{260} units. The fractions containing the tRNA are then pooled, brought to pH 10 with NH_4OH and incubated for 20 min at 37° to discharge any amino acid esterified to the tRNA.

4

The pH is brought back to neutrality and the tRNA is precipitated with potassium acetate-ethanol as before. The final precipitate is redissolved in water, dialysed extensively against water and lyophilized. With this procedure we have obtained 690 mg of tRNA from 800 g of wheat germ. This preparation accepted 1·2 $\mu\mu$moles of methionine and 0·8 $\mu\mu$moles of serine per 100 $\mu\mu$moles of tRNA. Glitz and Dekker (1963) have also described a somewhat similar method for the preparation of tRNA from wheat germ. Previous to the homogenization with phenol, these investigators extract the tissue with 5 volumes of an ethanol–ether (1 : 1) mixture to de-fat the material. This step is advisable with some plant materials such as peanut cotyledons and corn seeds that have a high content of oil.

B. Preparation of Aminoacyl-tRNA

For the study of the last steps of protein synthesis and for the purification of specific tRNAs, it is necessary to prepare and isolate radioactive aminoacyl-tRNAs. The formation of aminoacyl-tRNA can be accomplished by the method described in Section III as an assay system for the aminoacyl-tRNA synthetases. In order to conserve isotope the larger preparations are made with concentrations of up to 3 mg/ml of tRNA. Higher concentrations inhibit the reaction. The time of incubation is also increased to 40 min and excess enzyme is used. If a crude 105,000g fraction is used as a source of enzyme it should be dialysed or passed through Sephadex to eliminate contaminating amino acids. In general, it is better to use an enzyme preparation that has been passed through DEAE-cellulose as described by Muench and Berg (1966) to eliminate the tRNAs present in the preparations.

At the end of the incubation, an equal volume of water-saturated phenol is added and the mixture is shaken thoroughly. The water layer is separated by centrifuging and subsequently the water-soluble aminoacyl-tRNA is precipitated by addition of one-tenth of a volume of 20% potassium acetate and 2·5 volumes of ethanol at $-20°$. After 2 h at $-20°$, the precipitate is collected by centrifuging, redissolved in a minimal volume of water and dialysed for 8 h against several changes of water at 4°. The radioactive aminoacyl-tRNA can then be lyophilized and stored as a dry powder at $-20°$. Under these conditions the aminoacyl-tRNA is very stable.

C. Fractionation of tRNA from Plants

Several methods have been described for the fractionation and purification of transfer RNAs. The most important contributions in this field have been obtained by the counter-current distribution technique (Holley et al., 1961) and by several column chromatographic methods (Sueoka and Yamane, 1962; Muench and Berg, 1966; Kelmers et al., 1965; Gillam et al., 1967). These methods have established the heterogeneity of the tRNAs specific for each amino acid. Some of the heterogeneity is due to the requirements of the adaptor function of the tRNAs because of the degeneracy of the genetic code (Weisblum et al., 1962; Marshall et al., 1967).

In our laboratory we have examined the chromatographic profiles on methylated serum albumin-kieselguhr columns (MAK) of nine aminoacyl-tRNAs from wheat embryos before and after 4 days of germination (Allende, 1968). In these experiments we have followed closely the technique of Sueoka and Yamane (1962). Figure 5 shows the pattern for two aminoacyl-tRNAs. In one case, that of glutamyl-tRNA, it would appear that germination with its concomitant burst of protein synthesizing activity does not result in the alteration of the chromatographic profile of this tRNA. In the case of valyl-tRNA, however, a relative change in the amount of the two chromatographic fractions is apparent.

The MAK columns are fairly reproducible and constitute a valuable analytical tool. However, they have the important limitation of capacity of charge. The column described above could fractionate a maximum of only 2 mg of aminoacyl-tRNA. For this reason, for preparative experiments the reverse phase chromatographic columns described by Kelmers et al. (1965) are more appropriate. Figure 6 shows the elution pattern of methionyl-tRNA obtained from germinated and ungerminated embryos fractionated on the same column by using different isotopes to label the amino acid. There appears to be a small relative difference between the two clearly separated peaks. Using a similar column, we fractionated 200 mg of ungerminated wheat methionyl-tRNA which enabled us to study the coding properties of the two methionine peaks. Figure 7 shows the highly heterogeneous pattern of seryl-tRNA obtained by phenol extraction of 23 g of tissue obtained by the in vitro culture of soy bean callus tissue according to Miller (1963). It is evident, therefore, that the reverse phase chromatographic procedure is an excellent tool for both preparative and analytical studies of plant tRNAs.

V. AMINOACYL TRANSFER ENZYMES

When ribosomal particles are repeatedly washed by resedimentation, they become unable to catalyse the transfer of the aminoacyl groups

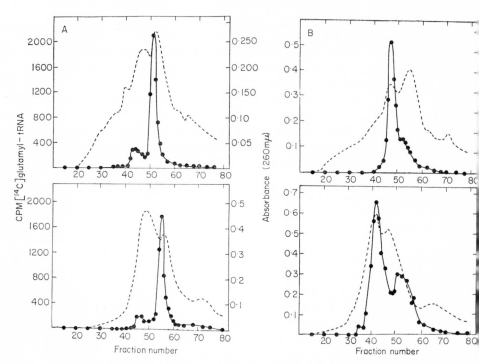

FIG. 5. Chromatography of aminoacyl-tRNAs of germinated and ungerminated wheat embryos on methylated serum albumin-kieselguhr columns. Approximately 1·3 mg of each [¹⁴C]aminoacyl-tRNA was applied to a MAK column (4·6 × 2·8 cm) prepared as described by Sueoka and Yamane (1962). The column was eluted with a linear salt gradient from 0·2M to 1·1M NaCl in 0·05M phosphate buffer, pH 6·3. Each gradient chamber contained 110 ml. The flow was regulated to 0·5 ml/min, collecting fractions of 2·0 ml. The optical density at 260 mμ was recorded, 200 μg of carrier RNA were added to each fraction and the nucleic acid was precipitated by adding one-tenth volume of 50% trichloroacetic acid. After 10 min in the cold, the samples were filtered on Millipore membranes which were subsequently dried and counted.

A. [¹⁴C]Glutamyl-tRNA from wheat embryos before germination (top) and after 4 days of germination (bottom). The two tRNA preparations were prepared by acylation with enzyme obtained from wheat embryos in the same state of germination.

B. [¹⁴C]Valyl-tRNA from germinated (top) and ungerminated (bottom) embryos. ●—● Indicates CPM; – – – indicates absorbance 260 mμ.

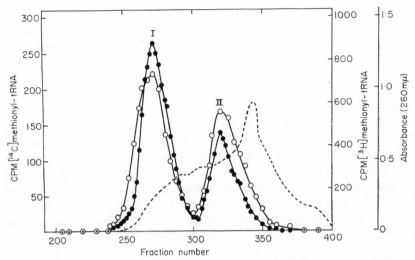

FIG. 6. Reverse-phase chromatography of methionyl-tRNA from germinated and ungerminated wheat embryos. Approximately 40 mg of germinated [^{14}C]-methionyl-tRNA and 36 mg of ungerminated [^{3}H]methionyl-tRNA were fractionated in the same chromatographic column and under the same conditions as described in Fig. 4. The absorbance at 260 mμ (– – –) was measured in a recorder and the radioactivity ^{3}H (●—●, ungerminated embryos) and ^{14}C (○—○, germinated embryos) determined in the counting mixture described by Bray (1960) in a scintillation spectrometer.

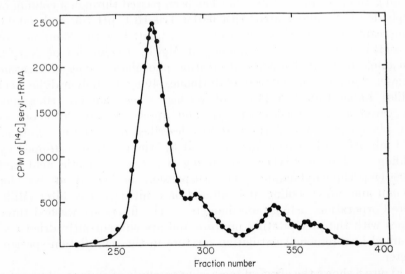

FIG. 7. Reverse-phase chromatography of [^{14}C]seryl-tRNA from soy bean callus tissue. Approximately 2·5 mg of [^{14}C]seryl-tRNA (84,000 CPM), obtained from soy bean callus tissue grown in culture in the presence of N^{6}-isopentenyl-adenosine, were chromatographed in the same reverse-phase chromatographic column as described in Fig. 4. In this case, however, the salt gradient was from 0·40M to 0·95M NaCl. To obtain the chromatographic profile, the entire fractions were precipitated by adding trichloroacetic acid to a final concentration of 5% at 4°. The precipitated tRNA was retained on Millipore membranes which were dried and counted. (Unpublished results, C. Allende.)

from the aminoacyl-tRNAs to the nascent polypeptides in the presence of messenger RNA. This deficiency can be overcome by the addition of enzymic factors called aminoacyl transfer enzymes or aminoacyl polymerization factors. These factors are present in the high-speed supernatant fluid of cell extracts.

The aminoacyl transfer enzymes were first described in mammalian tissue (Takanami and Okamoto, 1960; Grossi and Moldave, 1960) and soon after in *E. coli* (Nathans and Lipmann, 1961). Subsequent work (Fessenden and Moldave, 1962; Allende *et al.*, 1964; Lucas-Lenard and Lipmann, 1966) showed that there were several enzymes involved in each system. Transfer enzymes have since been detected in plant systems (Mans and Novelli, 1961; Allende and Bravo, 1966; Parisi *et al.*, 1967) and some of their properties have been studied. The preparation of the high-speed supernatant fluid of wheat embryos used below is described in the method for preparing ribosomes.

A. Assay

The assay for aminoacyl transfer enzymes which we have used in a wheat embryo system is as follows. The enzyme source is 50 μl of a high-speed supernatant fraction from wheat embryo extract, containing 15 mg of protein/ml, that has been passed through a column of Sephadex G-25 and eluted with 0·01M Tris-HCl, pH 7·5, and 0·001M 2-mercaptoethanol. The enzyme is incubated with 20 μmoles of Tris-HCl buffer, pH 7·5, 8 μmoles of MgCl$_2$, 100 μmoles of NH$_4$Cl, 0·5 μmole of GTP, 5 μmoles of creatine phosphate, 50 μg of creatine kinase, 2 μmoles of reduced glutathione, 40 μg of polyuridylic acid (Miles Laboratories), 0·14 mg of washed wheat embryo ribosomes (prepared as described on p. 57) and approximately 2000 cpm of [^{14}C]phenylalanyl-tRNA (prepared as described on p. 68) in a volume of 1 ml. After incubating for 15 min at 37° the reaction is stopped by adding 4 ml of a mixture containing 5% trichloroacetic acid and 0·0001M [^{12}C]phenylalanine. The suspension is heated at 90° for 15 min and, after cooling, it is filtered on a nitrocellulose filter (Millipore Corporation, 0·45 μ pore diameter). The filters are washed three times with the precipitating mixture and are subsequently dried and counted in a toluene scintillation mixture in the scintillation spectrometer.

Figure 8 shows the effect of increasing amounts of aminoacyl transfer enzyme on the polymerization of phenylalanine.

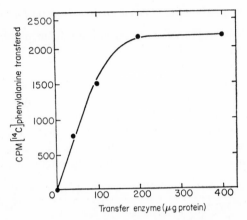

FIG. 8. The effect of wheat transfer enzyme concentration on the formation of polyphenylalanine from phenylalanyl-tRNA. The assay was carried out as described in the text. The enzyme preparation was a fraction from a high-speed supernatant fraction purified by precipitation between 40–80% saturation of ammonium sulphate. Phenylalanyl-tRNA (50 μg) was labelled with [^{14}C]phenylalanine (specific activity 100 μc/μmole) and contained approximately 3000 CPM.

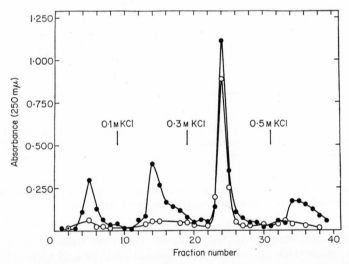

FIG. 9. DEAE-cellulose chromatography of wheat embryo transfer enzyme. Approximately 10 mg of protein, obtained by ammonium sulphate precipitation of a high-speed supernatant fraction of wheat embryo extract, were applied to a DEAE-cellulose column (10 × 0·9 cm) which had been equilibrated with 0·01M Tris-HCl, pH 7·5. Step-wise elution was achieved using the same buffer containing initially, no salt, and subsequently 0·1M, 0·3M and 0·5M KCl as indicated in the figure. The transfer activity of 50 μl aliquots was measured as described in the text. ● Indicates absorbance 280 mμ; ○ indicates transfer activity.

B. Properties of the Wheat Embryo Transfer Enzymes

The wheat embryo transfer enzymes have been partially purified by ammonium sulphate precipitation followed by DEAE-cellulose chromatography (Fig. 9). The protein fraction eluting with 0·3M KCl contains the transfer activity.

Further purification of the enzymes and attempts to resolve the activity into several complementary components, as has been done with enzymes from other sources, have been hindered by the extreme lability of these factors. Figure 10 shows the inactivation of the transfer enzymes at 37° and the stabilizing effect of glutathione.

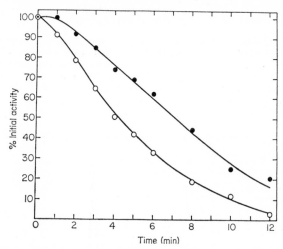

FIG. 10. Inactivation of the wheat embryo transfer enzymes. A partially purified preparation of wheat embryo transfer enzymes (400 μg) was incubated at 37° in a buffer containing 0·01M Tris-HCl, pH 7·5, in the presence of 0·02M glutathione (●—●) and in the absence of this compound (○—○). Aliquots were withdrawn at the times indicated and their aminoacyl transfer activity was determined as described in the text. The assay mixture contained in all cases 0·002M glutathione. (G. Oestreicher and J. E. Allende, unpublished results.)

It is interesting that the transfer enzymes from plant systems can function with ribosomes from animal sources and vice versa, but that these components cannot be exchanged with their bacterial counterparts (Table 2). Tobacco leaf chloroplast ribosomes, which were predominantly of the 66s variety, respond to the wheat embryo supernatant factors. Similar results have been obtained with the castor bean components in a fairly complete study of the species specificity of protein synthesis by Ciferri and his collaborators (Parisi *et al.*, 1967). These results, along with reports on the species specificity of yeast and insect transfer enzymes and ribosomes (Klink and Richter, 1966; Ilan and Lipmann, 1966), point to important differences between the components that intervene in protein synthesis in nucleated cells and in bacterial cells.

Further details of the aminoacyl transfer reaction are given in the section on *in vitro* protein synthesis systems.

TABLE 2

Species specificity of ribosomes and supernatant factors in the transfer reaction

Ribosomes	Supernatant fraction	cpm transferred
Wheat embryo	None	10
E. coli	None	14
Tobacco leaf chloroplast	None	60
Guinea pig liver	None	15
None	Wheat embryo	18
None	*E. coli*	20
None	Guinea pig liver	3
Wheat embryo	Wheat embryo	634
Wheat embryo	*E. coli*	24
Wheat embryo	Guinea pig liver	306
E. coli	Wheat embryo	29
E. coli	*E. coli*	796
E. coli	Guinea pig liver	15
Guinea pig liver	Wheat embryo	709
Guinea pig liver	*E. coli*	23
Guinea pig liver	Guinea pig liver	440
Tobacco leaf chloroplasts	Wheat embryo	530

The transfer assays were performed as described in the text, with 1800 cpm of [^{14}C]phenylalanyl-tRNA and 0·1 mg of RNA of the different ribosomal preparations. 2-Mercaptoethanol in a final concentration of 0·001M was present in the reaction mixture.

The supernatant fractions from *E. coli*, wheat embryo and guinea pig liver contained 1, 1·2 and 1·4 mg of protein, respectively.

C. A Protein that Interacts with GTP

Recently several laboratories have reported that aminoacyl transfer enzymes interact with guanosine triphosphate to yield a protein-GTP complex which can be isolated by retention on Millipore membranes or by gel filtration on Sephadex columns (Allende *et al.*, 1967; Gordon, 1967; Rao and Moldave, 1967).

In preliminary experiments we have found that the wheat embryo supernatant fraction also has a protein fraction which tightly binds GTP in the cold to form a complex which can be retained on Millipore filters (Table 3). This binding, which is specific for GTP, can also be detected by gel filtration on Sephadex columns.

TABLE 3

Interaction of GTP with a protein in the wheat embryo supernatant fraction

Experiment	Radioactive nucleotide	Additions	$\mu\mu$moles of bound nucleotide
1	GTP	— (signifies no addition)	27
	GTP	DNase (20 μg)	25
	GTP	RNase (25 μg)	21
	GTP	EDTA (0·025M)	6
	GTP	Enzyme heated 5 min at 60°	0·5
2	GTP	—	35·7
	ATP	—	14·6
	CTP	—	1·5
	UTP	—	0·7
3	GTP	—	42·4
	GDP	—	45·6
	GMP	—	4·7
4	GTP	—	23
	GTP	ATP (2·0 mμmoles) (non-radioactive)	23·4

The binding of GTP and other nucleotides was tested with 400 μg of a protein fraction of the 105,000*g* wheat embryo supernatant that precipitates from 40–80% saturation of ammonium sulphate.

The incubation mixture contained in a final volume of 200 μl: 0·05M Tris-HCl, pH 7·5, 0·05M NH$_4$Cl, 0·01M magnesium acetate, 1 mμmole of the radioactive nucleotide and 400 μg of the protein preparation. The incubation was for 5 min at 0°. The reaction was stopped by dilution with 5 ml of the same cold buffer–salt solution and the mixture was poured through a Millipore membrane (0·45 μ pore diameter). The membrane was washed three times with the same cold buffer, dried and counted in a liquid scintillation counter.

Several lines of evidence suggest that the GTP-binding protein and the transfer enzymes of wheat embryos are related. The protein responsible for the interaction with GTP fractionates together with the transfer enzymes through the ammonium sulphate and DEAE cellulose steps. The addition of aminoacyl-tRNA inhibits the retention of the complex on Millipore filters, but actually stabilizes the complex as can be observed by isolation on a Sephadex column. In this respect the GTP complex from wheat behaves similarly to that found in *E. coli* (Gordon, 1967).

VI. IN VITRO PROTEIN SYNTHESIS SYSTEMS DERIVED FROM PLANTS

The first cell-free system derived from plants that was reported to incorporate amino acids into protein was described by Stephenson *et al.* (1956) who worked with tobacco leaf extracts. Following this publication, a large number of systems from different sources have been reported (Rabson and Novelli, 1960; Mans and Novelli, 1961; Birnstiel *et al.*, 1961; Mikulska *et al.*, 1962; Heber, 1962; App and Jagendorf, 1963; Flamm *et al.*, 1963; Morton and Raison, 1963; Das *et al.*, 1964; Marcus and Feeley, 1964; Moustafa, 1964; Parisi and Ciferri, 1966; Allende and Bravo, 1966; Ellis and MacDonald, 1967; Leaver and Key, 1967). In all these cases, protein synthesis was measured by incorporation of radioactive amino acids into hot trichloroacetic acid precipitable material. The reports of Webster (1959b) and Raacke (1959) of net protein synthesis have not been reproduced by other workers (Lett and Takahashi, 1962).

One case has been reported in which a plant system, obtained from *Euglena gracilis*, was able to synthesize a specific protein, the coat protein of bacteriophage f2, when the virial RNA was supplied as genetic message (Schwartz *et al.*, 1965).

A. Cytoplasmic Amino Acid Incorporating Systems

The study of amino acid incorporating systems derived from the cytoplasm of plant cells has been greatly simplified by the work done previously with animal and bacterial systems since the essential features have been found to be similar in all species tested.

1. Requirements of the Systems

Three specific objectives can be pursued when carrying out *in vitro* amino acid incorporation experiments: (a) the determination of the relative activity of the protein synthesizing machinery of tissues under specific conditions. The obvious disadvantages of using cell-free systems for these studies are counterbalanced by the elimination of factors such as permeability barriers, cellular pools, etc. (b) The study of the properties of the components that intervene in the process of protein synthesis. (c) The achievement of the synthesis of a specific protein.

The pursuit of the first purpose requires that the investigator attempts to develop a system that tampers as little as possible with the components involved. These systems function, by definition, under direction of the endogenous messenger RNA and, therefore, are especially vulnerable to nucleases.

An example of the requirements of such a system may be quoted from the data of Parisi and Ciferri (1966) for a system obtained from castor bean seedlings (Table 4). The presence of ribosomes, supernatant fraction, ATP and its regenerating systems is absolutely

TABLE 4

Characteristics of the incorporation of L[¹⁴C]lysine into protein by the cell-free system from castor bean seedlings

Additions or omissions	$\mu\mu$moles [¹⁴C]lysine incorporated per reaction mixture
Complete system	16·01
− tRNA	11·15
− 105,000g supernatant	0·66
− Ribosomes	0·07
− 105,000g supernatant and tRNA	0·76
− Ribosomes and tRNA	0·09
− ATP, PEP, pyruvate kinase and GTP	0·41
− ATP, PEP and pyruvate kinase	0·46
− ATP	2·45
− GTP	12·23
− 19[¹²C]amino acids	10·98
+ RNAase (30 μg)	0·17
+ DNAase (5 μg)	13·80
Complete, deproteinized at 0 time	0·04

The incorporation of [¹⁴C]lysine (specific activity 180 μc/μmole) was carried out in 1·0 ml of a mixture containing: 100 μmoles Tris-HCl, pH 7·8, 10 μmoles magnesium acetate, 60 μmoles KCl, 3 μmoles ATP, 0·1 μmole GTP, 5 μmoles phosphoenolpyruvate, 20 μg pyruvate kinase, 15 μmoles 2-mercaptoethanol, 0·19 mg of RNA as washed ribosomes, 0·61 mg protein of a 105,000g supernatant fraction, 0·15 mg B. subtilis tRNA, 0·05 μmole of each of the remaining amino acids and 1·5–4·3 mμmoles (0·5 μc) of L[¹⁴C]lysine. The incubation was for 30 min at 37°. Reprinted with permission from Parisi and Ciferri (1966).

required. The other participants in the reaction, GTP, tRNA and 19 unlabelled amino acids, are present as contaminants of the ribosomes and supernatant fractions, and thus their absence results only in slight inhibition of the incorporation. These systems have been useful in the studies of the effect of seed germination and of light on the protein synthetic activity of plant tissues. These matters will be discussed in the following section.

If the amino acid incorporating system has been developed to achieve the synthesis of a specific protein produced by the particular source tissue, care must be taken to maintain the integrity of the system. The synthesis of specific animal proteins has been achieved *in vitro* in several animal systems (Bishop *et al.*, 1960; Campbell and Kernot, 1962) but the synthesis of a specific plant protein has not yet been achieved *in vitro* with plant systems. However, as noted above, Schwartz *et al.* (1965) have produced the synthesis of a bacteriophage

TABLE 5

Requirements for phenylalanine incorporation in a wheat embryo system from ungerminated seeds

Incubation system	[^{14}C]phenylalanine incorporated per reaction mixture	Activity
	cpm	%
Experiment 1		
Complete system	1935	100
Without ribosomes	13	1
Without poly U	62	3
Without tRNA	498	25
Without ATP	693	36
Without ATP and regenerating system	14	1
Without magnesium	16	1
Without GSH	2034	105
With puromycin (0·5 mM)	325	17
With chloramphenicol (0·1 mg)	1714	89
With RNase (50 μg)	85	4
Experiment 2		
Complete system	4040	100
Without supernatant fraction	720	18
Without GTP	1470	36
Without [^{12}C]amino acid mix	3850	96
Zero time	71	1

coat protein when a *Euglena gracilis* chloroplast system was purified to reduce the endogenous incorporation and to enhance the effect of the added viral messenger RNA.

If the purpose of the work is to study the properties of the components involved in the process of protein synthesis, a simplified system with isolated components and the best attainable incorporation is preferred. Such a system usually makes use of polyuridylic acid to direct the synthesis of polyphenylalanine as discovered by Nirenberg and Matthaei (1961).

The properties of a system isolated from ungerminated wheat embryos are shown in Table 5. It can be observed that the requirements for the addition of tRNA and GTP are more pronounced. As

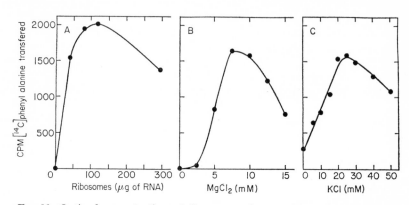

Fig. 11. Optimal concentrations of ribosomes and magnesium and potassium ions for amino acid incorporation in a wheat embryo system. The incorporation of [14C]phenylalanine directed by poly U was measured as described in Table 5. In A the effect of ribosomes, in B the effect of magnesium chloride and in C the effect of potassium chloride was measured.

expected, the absence of the other amino acids does not influence phenylalanine incorporation.

The optimal concentrations of Mg^{++}, K^+ and ribosomes for phenylalanine synthesis are shown in Fig. 11.

2. Genetic Code Determination

One very useful feature of the system derived from wheat embryos before germination is its extremely low endogenous amino acid incorporating activity. This property was exploited to determine the genetic code of higher plants (Basilio et al., 1966). Previously Sager et al. (1963) had found that an amino acid incorporating system from *Chlamydomonas* had a response to synthetic polynucleotides similar to that of *E. coli*.

Using the methods developed in the laboratories of Nirenberg and Ochoa (Nirenberg et al., 1963; Speyer et al., 1963), heteropolynucleotides with different base compositions were added to the wheat embryo system. The effect of the addition of the synthetic messages on the incorporation of each of 17 different amino acids was studied (Table 6).

TABLE 6

Amino Acid Incorporation in a Wheat Germ System Directed by Various Polynucleotides

	Polynucleotides and experiment number								Codons assigned
	None				UC (3:1)	UG (5:1)	UA (5:1)	UAG	
	1	2	3	4	1	2	3	4	
	μμmoles incorporated/ mg ribosomal RNA								
Phenylalanine	10·0	16·0	25·0	16·0	432·0	315·0	315·0	163·5	UUU
Leucine	18·4	18·3	26·5	28·3	235·6	90·6	77·5	85·0	2U1C, 2U1G, 2U1A, 2C1U
Proline	8·0	3·2	6·6		88·5	0·0	0·6		2C1U
Serine	7·3	2·7	11·4		91·9	0·3	0·0		2C1U
Alanine	8·2	2·2	9·3		0·0	0·2	0·0		
Glycine	6·8	4·1	7·8		0·4	9·9	0·4		2G1U
Isoleucine	9·6	3·8	12·0	14·0	0·0	0·0	55·4	30·6	2U1A
Arginine	5·7	2·7	6·4		0·8	1·4	0·0		
Glutamic acid	9·3	3·2	4·2	6·4	0·0	0·0	2·1	11·0	1U1A1G
Aspartic acid	4·4	2·1	3·8	19·1	1·2	0·2	6·3	15·2	1U1A1G
Lysine	7·4	2·9	32·0		0·0	0·2	19·0		2A1U
Histidine	10·1	3·2	6·6		0·0	0·0	0·0		
Tyrosine	8·3	4·2	10·5	12·6	0·0	0·0	49·5	39·5	2U1A
Methionine	7·4	2·9	9·2	1·2	0·0	0·1	0·0	9·3	1U1A1G
Tryptophan	5·5	7·9	4·4		1·7	22·1	0·0		2G1U
Valine	6·7	4·7	9·2		0·0	182·3	0·0		2U1G, 2G1U
Threonine	9·2	3·7	6·3		0·0	0·0	2·1		

For the assignment of the possible codewords for each amino acid, the relative frequency of the different triplets present in the random copolymer was compared to the relative incorporation of different amino acids influenced by that particular polynucleotide. The fitting of the data allowed the assignment of 17 codons out of the possible 64, showing a remarkable coincidence between the codons of higher plants with those determined for *E. coli* (Nirenberg *et al.*, 1966).

The results obtained with this approach, however, only allow one to arrive at the nucleotide composition of each triplet and not at its specific sequence. Nirenberg and Leder (1964) have devised a method for determining the exact sequences of the codons of each amino acid. This method makes use of the fact that aminoacyl-tRNAs bind to ribosomes in the presence of messenger RNA or trinucleotides that have the specific codons for that particular aminoacyl-tRNA. This ternary complex of ribosomes–messenger RNA–aminoacyl-tRNA can be isolated by retention on Millipore membranes. With this method, Nirenberg *et al.* (1966) have arrived at the specific sequence of all codons in *E. coli*.

Using the same method with washed ribosomes obtained from wheat embryos, binding of aminoacyl-tRNAs was directed by the presence of synthetic polynucleotides (Table 7).

TABLE 7

Binding of aminoacyl-tRNAs to ribosomes in presence of polynucleotides

[¹⁴C]aminoacyl-tRNA	Addition	Aminoacyl-tRNA, bound, $\mu\mu$moles	Δ $\mu\mu$moles
Phenylalanyl-tRNA			
	None	0·6	
	Poly U	10·7	10·1
	Poly U, GTP	6·7	6·1
	Poly U, supernatant	6·2	5·6
	Poly U, without Mg⁺⁺	0·5	0
	Poly C,	0·4	0
	Poly UC (3 : 1)	2·8	2·2
Prolyl-tRNA			
	None	0·4	
	Poly U	0·4	0
	Poly C	0·3	0
	Poly UC (3 : 1)	1·1	0·7
Histidyl-tRNA			
	None	0·2	
	Poly U	0·2	0
	Poly C	0·2	0
	Poly UC (3 : 1)	0·3	0·1

Under the conditions used, there did not seem to be a requirement for GTP or soluble enzymes for this binding. In the light of recent findings (Arlinghaus et al., 1964; Salas et al., 1967; Lucas-Lenard and Lipmann, 1967; Allende and Weissbach, 1967) it should be checked whether the binding of aminoacyl-tRNAs with substituted amino groups requires the presence of these factors at lower concentrations of Mg^{++}.

Fortunately, wheat embryo ribosomes seem to have an affinity for trinucleotides which resembles that of E. coli ribosomes and which allows one to carry out the binding assay with specific triplet codons (animal and yeast ribosomes do not seem to respond to triplets in the binding assay). During a recent stay in Dr Nirenberg's laboratory, we have checked the coding properties of wheat embryo methionyl-tRNA by the binding system. Table 8 shows the results obtained with the unfractionated methionyl-tRNA and with the two species obtained from reverse-phase chromatography as described in Section IV. It is evident that both ApUpG and GpUpG code for both methionyl-tRNA peaks. The response to UpUpG and CpUpG is negligible and poly UG seems to function only with the unfractionated tRNA.

It is interesting that the wheat ribosomes are not as effective as E. coli ribosomes in binding their own aminoacyl-tRNA and also have a much lower non-specific binding of aminoacyl-tRNA in the absence of the triplet.

Attempts to formylate the methionyl-tRNA fraction with either wheat embryo extract or the purified E. coli transformylase (Dickerman et al., 1967) were not successful. It would be very interesting, of course, to determine whether a substituted methionyl-tRNA plays the same important initiator function in plants that it seems to do in bacteria (Adams and Capecchi, 1966; Webster et al., 1966). In this respect, it may be relevant to remember that the Euglena gracilis chloroplast system was able to synthesize f2 phage coat protein (Schwartz et al., 1965) which has been shown to contain a formylmethionine end group.

A. Binding assay

The binding assay with plant components can be performed using the procedure of Nirenberg and Leder as follows. Two A_{260} units of washed ribosomes are incubated with 0·15 A_{260} units of the trinucleotides or 0·25 A_{260} unit of polynucleotide, and with approximately 0·5 A_{260} unit of the [^{14}C]aminoacyl-tRNA (containing at least 500 cpm) in a mixture containing 0·02M magnesium acetate, 0·05M potassium acetate and 0·05M Tris-acetate buffer, pH 7·2. The total volume is 50 μl. After 15 min incubation at 25°, the reaction is diluted by adding

3 or 4 ml of a cold solution of 0·05M potassium acetate, 0·05M Tris-
acetate, pH 7·2, and 0·02M magnesium acetate. The solution is passed
through a Millipore filter (0·45 μ pore diameter) and the filter is

TABLE 8

Binding of wheat methionyl-tRNA to ribosomes in the presence of different
oligonucleotides

Methionyl-tRNA	Ribosomes	Oligo-nucleotides	[^{14}C]methionyl-tRNA, bound μμmoles	Δ μμmoles
Unfractionated	Wheat	None	0·08	
		ApUpG	0·50	0·42
		GpUpG	0·11	0·03
		UpUpG	0·18	0·10
		Poly UG (3 : 1)	0·32	0·24
	E. coli	None	0·29	
		ApUpG	1·47	1·18
		GpUpG	0·67	0·38
		UpUpG	0·52	0·23
		Poly UG (3 : 1)	0·48	0·18
Methionyl-tRNA I	E. coli	None	0·23	
		ApUpG	0·70	0·47
		GpUpG	0·34	0·11
		UpUpG	0·27	0·04
		CpUpG	0·18	0·05
		UpUpU	0·22	0·01
		Poly UG (3 : 1)	0·27	0·04
Methionyl-t-RNA II	E. coli	None	0·06	
		ApUpG	0·53	0·47
		GpUpG	0·20	0·14
		UpUpG	0·07	0·01
		CpUpG	0·08	0·02
		Poly UG (3 : 1)	0·10	0·04

The binding of wheat embryo [^{14}C]methionyl-tRNA (acylated with wheat embryo
enzyme) was done exactly as described in the text. The oligonucleotides were kindly
supplied by Dr Marshall Nirenberg.
Methionyl-tRNA I and II were obtained by the reverse-phase chromatographic
procedure that fractionates wheat methionyl-tRNA into two main fractions (see Fig. 6).

thoroughly washed with the same buffer. The amount of aminoacyl-
tRNA bound to the ribosomes is determined by counting the dried
filters in the scintillation counter.

A control experiment without added oligonucleotide is always included, since there is some non-specific binding of aminoacyl-tRNA to ribosomes. The results should be expressed as the increased amount of aminoacyl-tRNA bound, due to the addition of the oligonucleotide. When using polynucleotides it is important to perform control experiments in the absence of ribosomes, since some binding of aminoacyl-tRNA to polynucleotides has been detected by this assay (M. W. Nirenberg, personal communication). This phenomenon does not occur when triplets are used.

3. Aminoacyl Transfer from Aminoacyl-tRNA to Nascent Polypeptide

For further investigation of the process of protein synthesis, one can resort to the simplest system available for polypeptide formation, the "transfer" system. As mentioned in Section V, this system can synthesize proteins (polypeptides) from isolated aminoacyl-tRNAs in the presence of washed ribosomes, a polynucleotide acting as synthetic messenger, partially purified transfer enzymes, GTP, SH groups and salts (Table 9). With this purified system the requirement for GTP is

TABLE 9

Requirements of the aminoacyl-tRNA transfer system from wheat embryos

System	cpm transferred per reaction mixture	Activity
		%
Complete	294	100
Without ribosomes	8	3
Without supernatant fraction	28	9
Without poly U	14	5
Without magnesium	5	2
Without KCl	56	19
Without GTP	30	10
Without phosphoenolpyruvate	40	13
Without pyruvate kinase	87	29
Without phosphoenolpyruvate and pyruvate kinase	16	5

The transfer system has been described in the text. Washed ribosomes and [^{14}C]phenylalanyl-tRNA (specific activity 100 μc/μmole) from rat liver were used in the assay.

absolute. Undoubtedly several reactions are involved in the transfer process, but as yet the nature and mechanism of these reactions are unknown. Figure 12 shows the optimal concentration of ribosomes,

Mg^{++} and K^+ for aminoacyl transfer in a wheat embryo system. It is interesting to note that the requirement for the monovalent cation K^+ (or NH_4^+) is absolute and that an optimum is obtained at the considerable concentration of 100 μmoles/ml. In this respect the system differs from the wheat amino acid incorporating system (see Fig. 11), but resembles the *E. coli* transfer system (Nakamoto *et al.*, 1963).

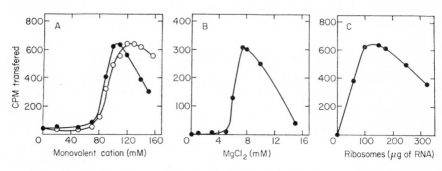

FIG. 12. Optimal concentrations of ribosomes, magnesium and monovalent cations for phenylalanyl transfer in a wheat embryo system. Phenylalanyl transfer was measured as described in Table 9. In A, potassium chloride (●—●) and ammonium chloride (○—○) were used as sources of monovalent cations. In B and C the effect of $MgCl_2$ and ribosomes, respectively, were tested.

B. Amino Acid Incorporating Systems Derived from Plant Organelles

Several *in vitro* systems for amino acid incorporation have been developed using the cellular organelles of plants.

1. Nuclei

Birnstiel and his collaborators (Birnstiel *et al.*, 1961; Birnstiel *et al.*, 1962; Flamm *et al.*, 1963) have presented evidence for amino acid incorporation in isolated nuclei from peas and from tobacco cell cultures. This work has emphasized the importance of isolating the nuclei free of contaminating cytoplasmic material which could affect the amino acid incorporation studies. The technique for isolating nuclei from cultures of tobacco cells derived from the pith of *Nicotiana tabacum*, var. *Xanthi*, used by these investigators (Flamm *et al.*, 1963), is briefly reproduced here.

Growing cells (5 days after inoculation of a fresh medium) were filtered through miracloth and homogenized with a glass-teflon homogenizer (two strokes at 200 rev/min) in half a volume of solution containing 0·25M sucrose and 0·01M Tris-HCl, pH 7·4. The homogenate was filtered through two layers of miracloth to separate out the cell walls. (Some nuclei remain with the walls and these can be recovered by resuspending this material in 2 volumes of grinding medium and forcing the suspension through a miracloth filter.) The filtrates are centrifuged at 200g for 20 min to yield a nuclear pellet. This pellet is resuspended in 1 volume of the same medium and centrifuged again at 200g for 20 min. The nuclei were further purified by centrifuging through a solution containing 18% dextran (mol. wt. 60,000–80,000, Nutritional Biochemical Corporation) in 0·25M sucrose and 0·0006M Tris with a final pH of 7·0. For this purpose the nuclei suspension was layered on the above solution and centrifuged for 15 min at 8000 rev/min in a SW-25 Spinco rotor. The pellet of purified nuclei was resuspended in 1 volume of grinding medium and used for the incubations with radioactive amino acids.

This nuclear preparation when incubated with DL-[^{14}C]lysine, incorporates isotope into hot, acid-insoluble material. This incorporation, which occurs only in the presence of added 0·25M sucrose and 0·01M Tris-HCl, pH 7·0, differs from that observed with cytoplasmic systems in that it is not inhibited by 0·01M EDTA, or by potassium ion (0·035M). The authors also indicate a difference with respect to the resistance of the nuclear system to large concentrations of chloramphenicol (2 mg/ml). The value of this criterion is doubtful in view of the conflicting reports on the effect of chloramphenicol on plant cytoplasmic systems (see the following section and Tables 5 and 11). Disruption of the nuclei with glass beads demonstrated that the bulk of the amino acid incorporated by these organelles was in the particulate fraction, and only 8% was present in the soluble phase of the nuclei.

2. Chloroplasts

Incorporation of amino acids into proteins by isolated chloroplasts has been widely reported (Heber, 1962; App and Jagendorf, 1963; Sissakian et al., 1963; Eisenstadt and Brawerman, 1964; Francki et al., 1965; Hall and Cocking, 1966). As with nuclei, one of the most important criteria in these investigations is the isolation of the intact organelle free from cytoplasmic contamination. The method used by Wildman's group (Francki et al., 1965) for the isolation of tobacco leaf chloroplasts is briefly described here. Five g fresh weight of leaves from young plants (Nicotiana glutinosa or Nicotiana tabacum) of 7 cm in height were de-veined and minced thoroughly in 6 ml of cold extracting medium: 2·5% Ficoll, 5% dextran, 25 mM Tris, pH 7·8, 1 mM magnesium acetate and 4 mM mercaptoethanol (Honda et al.,

1962). The resulting brew was filtered through three layers of fine gauze and the cellular extract was centrifuged at 1000g for 5–10 min. The pellet was resuspended in a medium containing 10 mM Tris, pH 7·8, 5 mM magnesium acetate and 3 mM mercaptoethanol. This fraction contained chloroplasts, nuclei and starch grains but it was considered that the incorporation of amino acids was due mainly to the plastids (Spencer and Wildman, 1964).

Dr C. R. Stocking (1959) has developed a non-aqueous system for the isolation of chloroplasts from leaves. This method apparently avoids the leaching out of plastid materials that occurs in aqueous media. The method detailed by Stocking et al. (1968) is briefly described here. Dry leaf material (0·4 mg) is ground at − 25° in 30 ml of a hexane-carbon tetrachloride mixture with a density of 1·26 at 20°. The grinding is carried out for 15 sec in a Virtus homogenizer with 8 g of glass homogenizing beads. After grinding, the preparation is filtered through cheese-cloth and an aliquot (5 ml) is layered on top of 22 ml of a linear density gradient (in a 30 ml capacity swinging bucket centrifuge tube) prepared with hexane and carbon tetrachloride. The density increases from 1·27 at the top of the tube to 1·49 at the bottom of the tube. After centrifuging for 2½ h at 35,000g, the fractions are collected. The upper green band contains the pure preparation of plastids.

TABLE 10

Effect of exogenous components on amino acid incorporation by the 1000g particulate fraction of tobacco leaves

Assay system	[14C]valine incorporated
	cpm
Complete reaction mixture	454
Omit GTP, UTP and CTP	417
Omit [12C]amino acids	343
Omit PEP and pyruvate kinase	290
Omit ATP, PEP and pyruvate	66
Omit all additions	24

The complete reaction mixture included: 0·4 ml of leaf extract in Honda (Honda et al., 1963) medium, 1000g fraction together with Tris, pH 7·8 (2·3 μmoles), magnesium acetate (2·3 μmoles), KCl (13·8 μmoles), mercaptoethanol (1·3 μmoles), ATP (0·2 μmole), pyruvate kinase (10 μg), UTP (0·01 μmole), CTP (0·01 μmole), GTP (0·01 μmole), a mixture of 0·0125 μmole of each of 19 amino acids (omitting valine) and [14C]valine (2·5 mμmoles and 0·5 μc) in a total volume of 0·5 ml. The incubation was carried out for 45 min at 25°. The reaction was stopped by addition of [12C]valine (50 μmoles) and 5% cold trichloroacetic, and the precipitate was washed and filtered before counting the radioactivity. Reproduced from Spencer and Wildman (1964) with permission.

Ribosomes obtained from these chloroplasts were shown to be active in an aminoacyl transfer system (Allende and Bravo, 1966). There are some conflicting reports regarding the requirements of isolated chloroplasts for amino acid incorporation. Thus App and Jagendorf (1963) found no stimulation of the activity upon addition of ATP (1 mM) and an ATP generating system. Hall and Cocking (1966) find only a slight stimulation while Spencer and Wildman (1964) and Eisenstadt and Brawerman (1963) find a definite requirement for these components. The effect of ribonuclease is also unclear and apparently depends on the method of chloroplast preparation and the source of the organelles. All authors, however, seem to agree that both chloramphenicol and puromycin inhibit the chloroplast systems.

As discussed in the section on ribosomes, it is clear that chloroplasts contain ribosomes and that the incorporation of amino acids occurring in plastids occurs in all probability in the plastid ribosomal particles (Boardman et al., 1965).

The amino acid incorporating systems with ribosomes isolated from chloroplasts seem to show all the requirements and properties of the cytoplasmic systems (see Table 10).

3. Mitochondria

Das and his collaborators (Das and Roy, 1961; Das et al., 1964; Chatterjee et al., 1966) have reported incorporation of amino acids into protein of the isolated mitochondria from Vigna sinencis (Linn.). These authors observed that the incorporation of glutamate using seedlings from these plants was 25 to 30 times greater in the mitochondrial fraction than in the microsomal fraction. Other peculiarities of the system were that ATP or GTP had little stimulating effect, and that the incorporation velocity increased with time. Thus the incorporation after 2 h was ten times greater than the incorporation after 30 min. Incorporation of amino acids proceeded for over 6 h, reaching a net protein synthesis of approximately 50 μg/mg of original protein. Bacterial contamination of the incubation mixture (between 0.3×10^6 to 2×10^6 bacterial counts) was not considered to be contributing to the activity since penicillin G (100 units/ml) did not have any effect on the activity of the system. More evidence, however, ruling out bacterial contamination as a factor in these systems would be very desirable. Ellis and MacDonald (1968), working with sterile beet disks, have recently described amino acid incorporation in the mitochondrial fraction. This incorporation was reported to be energy dependent and resistant to ribonuclease and cycloheximide but can be inhibited by chloramphenicol.

C. Bacterial Contamination

One of the main concerns of investigators studying systems that incorporate amino acids into proteins should be whether their results can be attributed to bacterial contamination. This problem is especially bothersome in the study of incorporation by organelles since these structures have a size similar to bacteria and sediment at similar velocities. Since even under careful aseptic conditions, bacteria have been found in the components that intervene in incorporating systems (Hall and Cocking, 1966), the important question is whether the amount of bacteria present can explain the results or can significantly contribute to them. The following criteria may be useful in determining whether bacterial contamination is a problem.

1. Kinetics of Incorporation

A study of the kinetics of amino acid incorporation gives possibly the best clue. If the incorporation activity continues at a good rate after $1\frac{1}{2}$ h or 2 h, bacterial contamination should be suspected. All cell-free systems are usually exhausted by that time and need complementation with energy sources, messenger RNA or even with fresh enzymes.

2. Requirements of the System

Since wild-type bacteria have minimal requirements, the more ingredients that are found to be necessary for protein synthesis, the less likely that the activity is due to bacterial growth. Here again the organelle systems are harder to differentiate since they contain more endogenous material.

3. Antibiotics and Inhibitors of Protein Synthesis

Some inhibitors of protein synthesis act preferentially on bacterial systems (chloramphenicol and streptomycin). Resistance to doses that are known to inhibit greatly bacterial systems is a good sign. Other antibiotics such as cycloheximide act preferentially on systems from nucleated cells. Susceptibility to these components argues, therefore, against bacterial contamination.

Other substances such as penicillin and sulfa drugs, which act on bacterial growth and do not interfere with the mechanism of protein synthesis, can be useful in maintaining a low level of bacterial contamination. It must be remembered, however, that even a combination of these antibiotics does not eliminate the whole spectrum of bacteria.

4. Identification of the Final Product

Identification of the product is undoubtedly the ultimate criterion. If the protein synthesized is easily identifiable such as hemoglobin in the reticulocyte system, the problem is eliminated. However, even such simple experiments as an analysis of the untreated product in a density gradient, or radioautography, can serve to rule out the possibility that contaminating bacteria are responsible for the incorporating activity.

D. Effect of Antibiotics on Plant Protein Synthesis Systems

The following antibiotics, which inhibit protein synthesis, have been tested in plant systems: puromycin, streptomycin, chloramphenicol, tetracyclines, cycloheximide, sparsomycin and actinomycin D. Puromycin is a potent inhibitor of protein synthesis in all species. It has been shown to act at the step that synthesizes the peptide bond (Allen and Zamecnik, 1962; Nathans, 1964). This antibiotic at a concentration of 10^{-4} M is effective in practically all plant systems tested (Table 11).

Streptomycin acts on bacterial ribosomes and causes a misreading of the genetic message (Davies et al., 1964). This antibiotic has been tested on plant systems (Marcus and Feeley, 1965; Parisi and Ciferri, 1966) and was found to have little effect. Chloramphenicol is a potent inhibitor of the transfer step of protein synthesis in bacterial systems, but very high concentrations are required to affect animal systems. This compound seems to inhibit chloroplast incorporating systems in vitro (App and Jagendorf, 1964; Spencer and Wildman, 1964), but little effect is observed in cytoplasmic systems from different sources (Marcus and Feeley, 1965; Parisi and Ciferri, 1966; Allende and Bravo, 1966; Ellis and MacDonald, 1967). Tetracyclines have been shown to inhibit protein synthesis by blocking the binding of aminoacyl-tRNA to the ribosome-messenger complex in bacterial systems (Suarez and Nathans, 1965; Hierowsky, 1965). Oestreicher, in our laboratory, has shown that different tetracycline derivatives also inhibit the binding of phenylalanyl-tRNA to a wheat ribosome-poly U complex (see Table 11). Chlortetracycline seems to be the most potent inhibitor. Sparsomycin, an antibiotic of unknown structure which has been shown to inhibit the transfer step of protein synthesis in E. coli (Goldberg and Mitzugi, 1967), is an inhibitor more potent than puromycin in the wheat embryo transfer system (see Table 11).

Cycloheximide, a glutarimide antibiotic which inhibits the transfer step of protein synthesis in yeast and mammalian cells (Siegel and

Sisler, 1965; Bennett et al., 1965), was found to be a potent inhibitor of the wheat amino acid incorporation system (Marcus and Feely, 1966). Actinomycin D inhibits protein synthesis by interacting with DNA and blocking the transcription process (Reich, 1964). This antibiotic, therefore, only acts on in vitro systems that involve the transcription

TABLE 11

Effect of some antibiotics on wheat protein synthesis systems

A   Amino acid incorporation system

Inhibitor	[¹⁴C]phenylalanine incorporated cpm	% Inhibition
None	1935	0
Puromycin (0·5 mM)	325	83
Chloramphenicol (0·1 mg)	1714	11

B   Aminoacyl transfer system

Inhibitor	[¹⁴C]phenylalanine transferred cpm	% Inhibition
None	820	0
Puromycin (0·5 mM)	115	86
Sparsomycin (0·05 mM)	16	98

C   Aminoacyl-tRNA binding system

Inhibitor	[¹⁴C]phenylalanyl-tRNA bound cpm	% Inhibition
None	1100	0
Tetracycline (0·1 mM)	535	52
Chlortetracycline (0·1 mM)	297	73
Dimethylchlortetracycline (0·1 mM)	418	62

The three assays used were performed exactly as described in the text. In all cases [¹⁴C]phenylalanine (specific activity 100 μc/μmole) and 40 μg of polyuridylic acid were used.

of DNA. It is interesting to note that the chloroplast system of Spencer and Wildman (1964) shows a small inhibition of amino acid incorporation by actinomycin D. The presence of DNA in chloroplasts seems to be well established (Lyttleton, 1962).

VII. THE REGULATION OF PROTEIN SYNTHESIS IN PLANTS

Knowledge about the process of protein synthesis amassed during the past decade has opened experimental approaches to some of the most fundamental problems of biology. Thus in the last few years biologists have been able to undertake research projects that have as goals the understanding of metabolic regulation, cellular differentiation, embryogenesis, and the mechanism of action of hormones at the molecular level. The present evidence clearly indicates that the regulation of protein synthesis plays a central role in all of these processes. Therefore, although we are still groping for better and more general ideas in this field, we will enumerate some of the experiments and techniques being used to attack these problems in plants.

A. Regulation of Transcription

The transcription step in protein synthesis entails the synthesis of a messenger RNA molecule with a base composition complementary to the DNA strand that carries the information for a particular protein. It is probable that the major regulation of protein synthesis occurs at this level.

Bonner et al. (1963) have performed some experiments that apparently demonstrate regulation of the synthesis of pea seed globulin at the transcription level. These workers isolated chromatin from developing pea cotyledons, tissues that actively synthesize globulin protein, and from pea vegetative buds, from which globulin is absent. After introducing these two chromatin preparations in an E. coli in vitro system that synthesized proteins but also required the presence of DNA (Wood and Berg, 1962), the synthesis of pea globulin was studied by immunochemical methods. A considerable amount of globulin (7% of all protein) was synthesized in the system containing cotyledon chromatin, while a negligible amount was present in the system with bud chromatin. If the bud chromatin, however, was denuded of histones with salt it acquired the property of producing pea globulin in the E. coli system. These experiments indicate that the process of cellular differentiation is accompanied by selective availability of the genetic material for transcription, and that histones might possibly play a role in this regulatory mechanism (but see Fambrough et al., 1968). Another example of regulation at the transcription level in plants is given by the effect of gibberellic acid on the induction of α-amylase in the aleurone layer of germinating barley seeds. Isolated barley aleurone layers do not produce the enzyme α-amylase after imbibition, unless the hormone gibberellic acid is supplied. In whole barley seeds the hormone is provided by the embryo. Varner and Chandra (1964) presented clear evidence that the induction of the enzyme in response to the hormone was mediated through de novo synthesis of the α-amylase. The appearance of α-amylase activity in the presence of gibberellic acid was inhibited by puromycin and actinomycin D, indicating that the regulatory function of the hormone was probably occurring at the transcription level. Johri and Varner (1968) have recently demonstrated that gibberellic acid stimulates the synthesis of RNA in isolated pea nuclei.

Leaver and Key (1967) and Ellis and MacDonald (1967) have studied the regulatory mechanisms that operate in disks excised from plant storage tissues. After the tissue is excised a great increment in the metabolic activity is observed. The work of these investigators presents evidence connecting the increase of protein synthesis with the *de novo* synthesis of messenger RNA and the formation of poly-ribosomes.

The enzyme that presumably carries out the transcription process, RNA polymerase, has been purified and studied from maize seedlings by Stout and Mans (1967).

B. Regulation of Translation

The translation process may be viewed as occurring in several steps: attachment of mRNA to ribosomes, formation of aminoacyl-tRNAs, initiation of the peptide chain, polymerization of amino acids, and termination and release of the completed protein. Each of the steps offers a possibility for regulation.

After water imbibition a plant seed changes rapidly from a dormant state with very little protein synthesis, to a state of rapid biosynthetic activity. The inactivity before germination is not due to any deficiency in the translation machinery (Marcus and Feeley, 1964; Allende and Bravo, 1966). Further evidence of Marcus and Feeley (1966) has shown that DNA transcription is not involved in the trigger mechanism of germination, but rather that the interaction of mRNA, already present in the dry seed but somehow "masked", with the ribosomes would constitute such a trigger.

Williams and Novelli (1968) have studied the effect of illumination of the whole plant on ribosomes of maize and bean seedlings grown in the dark. From their studies it appears that exposure to light, especially red light, of 5–7-day-old seedlings causes a stimulation of up to 200% in the activity of the ribosomes obtained from the shoots. The activity of the ribosomes for amino acid incorporation was tested *in vitro* in the presence of endogenous messenger RNA and also with added polyuridylic acid. Part of the increase of endogenous amino acid incorporating activity was apparently due to an increase in the amount of ribosomal particles occurring as polysomes. However, even in the presence of excess polyuridylic acid, the particles from the illuminated shoots were still more active in phenylalanine polymerization. There might, therefore, be a direct effect of light on the ribosomes or ribosome-bound enzymes in these tissues.

The availability of certain species of transfer RNAs required to translate specific messenger RNAs has been postulated as a mechanism for the regulation of protein synthesis (Ames and Hartman, 1963). In species other than bacteria, differences in the aminoacyl-tRNA content of cells under different physiological conditions have been found, suggesting a possible modulatory role (Kaneko and Doi, 1966; Lazzarini, 1966; Kano-Sueoka and Sueoka, 1966).

In our laboratory we have looked at the patterns of different amino-acyl-tRNAs in wheat embryos before and after 4 days of germination. Analysis by chromatography on methylated serum albumin columns showed some relative differences in the fractions of valyl-tRNA and methionyl-tRNA (see Figs 5 and 6). The other aminoacyl-tRNAs examined (those specific for threonine, phenylalanine, leucine, serine, aspartic acid, glutamic acid and glycine) did not show any appreciable differences (Allende, 1968).

A very interesting relationship between transfer RNA and plant hormone has been reported. Recent studies on the nucleotide sequence of transfer RNAs has established the presence of N^6-(γ, γ-dimethyl-allylamino) purine (N^6-isopentenyladenine) in a position next to the anticodon in seryl- and tyrosyl-tRNAs from yeast (Zachau et al., 1966; Hall et al., 1966). This compound and other N^6-substituted adenine derivatives have potent cytokinin activity (Beauchesne and Goutarel, 1963; Miller, 1961), inducing cell growth, division and differentiation. The incorporation of radioactive N^6-benzylaminopurine into the tRNA of soy bean tissue has been demonstrated by Fox and Chen (1967). Fittler and Hall (1966), using unfractionated yeast tRNA, have attempted the specific alteration of the isopentenyl moiety of seryl-tRNA by treatment with aqueous iodine solutions under mild conditions. After iodine treatment, the tRNA retains full capacity to accept serine, but the binding of the seryl-tRNA to ribosomes in the presence of poly UC is greatly reduced. The importance of the iso-pentenyladenine residue in the function of tRNA in protein synthesis remains uncertain, but the change in ribosome affinity upon alteration of this group suggests that it may affect the velocity of translation.

ACKNOWLEDGEMENTS

Some of the work presented in this chapter was done in our laboratory with the support of the Jane Coffin Childs Memorial Fund for

Medical Research. The following collaborators participated in this work: Catherine C. Allende, Carlos Basilio, Maria Bravo, Marta Gatica, Maria Matamala, Guillermo Oestreicher and German Zanghellini. I am grateful to Drs David Novelli, Rusty Mans, Orio Ciferri, R. J. Ellis and C. Ralph Stocking for making available unpublished material.

REFERENCES

Adams, J. M. and Capecchi, M. R. (1966). *Proc. natn. Acad. Sci. U.S.A.* **55**, 147–155.

Allen, D. and Zamecnik, P. C. (1962). *Biochim. biophys. Acta* **55**, 865–874.

Allende, J. E. (1968). *Natn. Cancer Inst. Monogr.* **27**, 169–179.

Allende, J. E. and Bravo, M. (1966). *J. biol. Chem.* **241**, 5813–5817.

Allende, J. E. and Weisbach, H. (1967). *Biochem. biophys. Res. Commun.* **28**, 82–88.

Allende, J. E., Monro, R. and Lipmann, F. (1964). *Proc. natn. Acad. Sci. U.S.A.* **51**, 1211–1216.

Allende, J. E., Seeds, N. W., Conway, T. W. and Weisbach, H. (1967). *Proc. natn. Acad. Sci. U.S.A.* **58**, 1566–1573.

Ames, B. N. and Hartman, P. E. (1963). *Cold Spring Harb. Symp. quant. Biol.* **28**, 349–353.

App, A. A. and Jagendorf, A. T. (1963). *Biochim. biophys. Acta* **76**, 286–292.

Arlinghaus, R. G., Shaeffer, J. and Schweet, R. S. (1964). *Proc. natn. Acad. Sci. U.S.A.* **51**, 1291–1299.

Attwood, M. M. and Cocking, E. C. (1965). *Biochem. J.* **96**, 616–625.

Basilio, C., Bravo, M. and Allende, J. E. (1966). *J. biol. Chem.* **241**, 1917–1919.

Bayley, S. T. (1964). *J. molec. Biol.* **8**, 231–238.

Beauchesne, B. and Goutarel, R. (1963). *Physiologia Pl.* **16**, 630–635.

Beinert, H., Green, D. E., Hift, H., van Korff, R. W. and Ramakrishnan, C. V. (1953). *J. biol. Chem.* **203**, 35–45.

Bennett, L. L., Ward, V. L. and Brockman, R. W. (1965). *Biochim. biophys. Acta* **103**, 478–485.

Berg, P. (1956). *J. biol. Chem.* **222**, 1025–1034.

Berg, P. and Ofengand, E. J. (1958). *Proc. natn. Acad. Sci. U.S.A.* **44**, 78–86.

Birnstiel, M. L., Chipchase, M. and Bonner, J. (1961). *Biochem. biophys. Res. Commun.* **6**, 161–166.

Birnstiel, M. L., Rho, J. H. and Chipchase, M. (1962). *Biochim. biophys. Acta* **55**, 734–740.

Bishop, J., Leahy, J. and Schweet, R. S. (1960). *Proc. natn. Acad. Sci. U.S.A.* **46**, 1030–1038.

Boardman, N. K., Francki, R. I. B. and Wildman, S. G. (1965). *Biochemistry, N.Y.* **4**, 872–876.

Boardman, N. K., Francki, R. I. B. and Wildman, S. G. (1966). *J. molec. Biol.* **17**, 470–489.

Bonner, J., Huang, R. C. and Gilden, R. (1963). *Proc. natn. Acad. Sci. U.S.A.* **50**, 893–900.

Brawerman, G. (1963). *Biochim. biophys. Acta* **72**, 317–331.

Bray, G. A. (1960). *Analyt. Biochem.* 1, 279–285.
Brunngraber, E. E. (1962). *Biochem. biophys. Res. Commun.* 8, 1–3.
Campbell, P. N. and Kernot, B. A. (1962). *Biochem. J.* 82, 262–266.
Chatterjee, S. K., Das, H. K. and Roy, S. C. (1966). *Biochim. biophys. Acta* 114, 349–354.
Clark, J. C., Jr. (1958). *J. biol. Chem.* 233, 421–424.
Clark, M. F., Matthews, R. E. F. and Ralph, R. K. (1964). *Biochim. biophys. Acta* 91, 289–304.
Das, H. K. and Roy, S. C. (1961). *Biochim. biophys. Acta* 53, 445–446.
Das, H. K., Chatterjee, S. K. and Roy, S. C. (1964). *J. biol. Chem.* 239, 1126–1133.
Davies, J. E., Gilbert, W. and Gorini, L. (1964). *Proc. natn. Acad. Sci. U.S.A.* 51, 883–890.
Davis, J. W. and Novelli, G. D. (1958). *Archs Biochem. Biophys.* 75, 299–308.
DeMoss, J. A. and Novelli, G. D. (1956). *Biochim. biophys. Acta* 22, 49–61.
Dickerman, H., Steers, E. Jr., Redfield, B. and Weisbach, H. (1967). *J. biol. Chem.* 242, 1522–1525.
Doctor, B. P. and Mudd, H. A. (1963). *J. biol. Chem.* 238, 3677–3681.
Dure, L. and Waters, L. (1965). *Science, N.Y.* 147, 410–412.
Eisenstadt, J. M. and Brawerman, G. (1963). *Biochim. biophys. Acta* 76, 319–321.
Eisenstadt, J. M. and Brawerman, G. (1964). *Biochim. biophys. Acta* 80, 463–472.
Ellis, R. J. and MacDonald, I. R. (1967). *Pl. Physiol., Lancaster* 42, 1297–1302.
Ellis, R. J. and MacDonald, I. R. (1968). *Pl. Physiol., Lancaster* (in press).
Fambrough, D. M., Fujimura, F. and Bonner, J. (1968). *Biochemistry, N.Y.* 7, 575–584.
Fessenden, J. M. and Moldave, K. (1962). *Biochemistry, N.Y.* 1, 485–490.
Fittler, F. and Hall, R. H. (1966). *Biochem. biophys. Res. Commun.* 25, 441–446.
Flamm, W. G. and Birnstiel, M. L. (1964). *Biochim. biophys. Acta* 87, 101–110.
Flamm, W. G., Birnstiel, M. L. and Filner, P. (1963). *Biochim. biophys. Acta* 76, 110–119.
Fox, E. J. and Chen, C.-M. (1967). *J. biol. Chem.* 242, 4490–4494.
Francki, R. I. B., Boardman, N. K. and Wildman, S. G. (1965). *Biochemistry, N.Y.* 4, 865–872.
Gillam, I., Millward, S., Blew, D., von Tigerstrom, M., Wimmer, E. and Tener, G. W. (1967). *Biochemistry, N.Y.* 6, 3043–3056.
Glitz, D. G. and Dekker, C. A. (1963). *Biochemistry, N.Y.* 2, 1185–1192.
Goldberg, I. H. and Mitzugi, K. (1967). *Biochemistry, N.Y.* 6, 372–383.
Gordon, J. (1967). *Proc. natn. Acad. Sci. U.S.A.* 58, 1574–1578.
Grossi, L. G. and Moldave, K. (1960). *J. biol. Chem.* 235, 2370–2374.
Hall, R. H., Robins, M. J., Stasiuk, L. and Thedford, R. (1966). *J. Am. chem. Soc.* 88, 2614–2615.
Hall, T. C. and Cocking, E. C. (1966). *Biochim. biophys. Acta* 123, 163–171.
Heber, U. (1962). *Nature, Lond.* 195, 91–92.
Hierowsky, M. (1965). *Proc. natn. Acad. Sci. U.S.A.* 53, 594–599.
Hoagland, M. B. (1955). *Biochim. biophys. Acta* 16, 288–289.
Hoagland, M. B., Stephenson, M. L., Scott, J. F., Hecht, L. I. and Zamecnik, P. C. (1958). *J. biol. Chem.* 231, 241–257.
Holley, R. W. (1965). *In,* "Plant Biochemistry" (J. Bonner and J. E. Varner, eds) pp. 346–360, Academic Press, New York and London.
Holley, R. W., Apgar, J., Doctor, B. P., Farrow, J., Marini, M. A. and Nerril, S. H. (1961). *J. biol. Chem.* 236, 200–202.

Honda, S. I., Hongladarom, T. and Wildman, S. G. (1963). *Pl. Physiol, Lancaster* **37**, XLI–XLII.

Ichimura, K., Izawa, M. and Oöta, Y. (1960). *Pl. Cell Physiol., Tokyo* **1**, 317–325.

Ilan, J. and Lipmann, F. (1966). *Acta biochim. pol. XIII*, 353–359.

Jacobson, K. B., Nichimura, S., Bernett, W. E., Mans, R. J., Cammarano, P. and Novelli, D. (1964). *Biochim. biophys. Acta* **91**, 305–312.

Johnston, F. B. and Stern, H. (1957). *Nature, Lond.* **179**, 160–161.

Johri, M. M. and Varner, J. E. (1968). *Proc. natl. Acad. Sci. U.S.A.* **59**, 269–276.

Kaneko, I. and Doi, R. (1966). *Proc. natn. Acad. Sci. U.S.A.* **55**, 564–571.

Kano-Sueoka, T. and Sueoka, N. (1966). *J. molec. Biol.* **20**, 183–209.

Kelmers, A. D., Novelli, G. D. and Stulberg, M. P. (1965). *J. biol. Chem.* **240**, 3979–3983.

Klink, F. and Richter, D. (1966). *Biochim. biophys. Acta* **114**, 431–433.

Lazzarini, R. A. (1966). *Proc. natn. Acad. Sci. U.S.A.* **56**, 185–190.

Lazzarini, R. A. and Mehler, A. H. (1966). *In*, "Procedures in Nucleic Acid Research" (G. L. Cantoni and D. R. Davis, eds) pp. 409–413, Harper and Row, New York.

Leaver, C. J. and Key, J. L. (1967). *Proc. natn. Acad. Sci. U.S.A.* **57**, 1338–1344.

Lett, J. T. and Takahashi, W. N. (1962). *Archs. Biochem. Biophys.* **96**, 569–574.

Littlefield, J. W., Keller E. B., Gross, J. and Zamecnik, P. C. (1955). *J. biol. Chem.* **217**, 111–123.

Lucas-Lenard, J. and Lipmann, F. (1966). *Proc. natn. Acad. Sci. U.S.A.* **55**, 1562–1566.

Lucas-Lenard, J. and Lipmann, F. (1967). *Proc. natn. Acad. Sci. U.S.A.* **57**, 1050–1057.

Lyttleton, J. W. (1962). *Expl. Cell Res.* **26**, 312–317.

Lyttleton, J. W. (1968). *Biochim. biophys. Acta* **154**, 145.

Mans, R. J. (1967). *A. Rev. Pl. Physiol.* **18**, 127–146.

Mans, R. J. and Novelli, G. D. (1961). *Biochim. biophys. Acta* **50**, 287–300.

Mans, R. J. and Novelli, G. D. (1964). *Biochim. biophys. Acta* **80**, 127–136.

Marcus, A. and Feeley, J. (1964). *Proc. natn. Acad. Sci. U.S.A.* **51**, 1075–1079.

Marcus, A. and Feeley, J. (1965). *J. biol. Chem.* **240**, 1675–1680.

Marcus, A. and Feeley, J. (1966). *Proc. natn. Acad. Sci. U.S.A.* **56**, 1770–1777.

Marshall, R. E., Caskey, C. T. and Nirenberg, M. (1967). *Science, N.Y.* **155**, 820–826.

Mikulska, E., Odintsova, M. S. and Sissakian, N. M. (1962). *Naturwissenschaften* **49**, 549.

Miller, C. (1961). *A. Rev. Pl. Physiol.* **12**, 395–408.

Miller, C. (1963). *In*, "Modern Methods in Plant Analysis" Vol. 6, pp. 194–202.

Monier, R., Stephenson, M. L. and Zamecnik, P. C. (1960). *Biochim. biophys. Acta* **43**, 1–8.

Morton, R. K. and Raison, J. K. (1963). *Nature, Lond.* **200**, 429–433.

Moustafa, E. (1963). *Biochim. biophys. Acta* **76**, 280–285.

Moustafa, E. (1964). *Biochim. biophys. Acta* **91**, 421–426.

Moustafa, E. and Lyttleton, J. W. (1963). *Biochim. biophys. Acta* **68**, 45–51.

Moustafa, E. and Proctor, M. H. (1962). *Biochim. biophys. Acta* **63**, 93–97.

Muench, K. and Berg, P. (1966). *In*, "Procedures in Nucleic Acid Research" (G. L. Cantoni and D. R. Davies, eds) pp. 375–383, Harper and Row, New York.

Nakamoto, T., Conway, T. W., Allende, J. E., Spyrides, G. J. and Lipmann, F. (1963). *Cold Spring Harb. Symp. quant. Biol.* **28**, 227–231.

Nathans, D. (1964). *Proc. natn. Acad. Sci. U.S.A.* **51**, 585–592.

Nathans, D. and Lipmann, F. (1961). *Proc. natn. Acad. Sci. U.S.A.* **7**, 497–504.

Nirenberg, M. W. and Leder, P. (1964). *Science, N.Y.* **145**, 1399–1407.

Nirenberg, M. W. and Matthaei, J. H. (1961). *Proc. natn. Acad. Sci. U.S.A.* **47**, 1588–1602.

Nirenberg, M. W., Jones, O. W., Leder, P., Clark, B. F. C., Sly, W. S. and Petska, S. (1963). *Cold Spring Harb. Symp. quant. Biol.* **28**, 549–557.

Nirenberg, M. W., Caskey, C. T., Marshall, R. E., Brimacombe, R., Kellogg, D., Doctor, B. P., Hatfield, D., Levin, J., Rottman, F., Petska, S., Wilcox, M. and Anderson, F. (1966). *Cold Spring Harb. Symp. quant. Biol.* **31**, 11–24.

Palade, G. E. (1955). *J. biophys. biochem. Cytol.* **1**, 59–68.

Parisi, B. and Ciferri, O. (1966). *Biochemistry, N.Y.* **5**, 1638–1645.

Parisi, B., Milanesi, G., van Etten, J. L., Perani, A. and Ciferri, O. (1967). *J. molec. Biol.* **28**, 295–309.

Raacke, I. D. (1959). *Biochim. biophys. Acta* **34**, 1–9.

Rabson, R. and Novelli, G. D. (1960). *Proc. natn. Acad. Sci. U.S.A.* **46**, 484–488.

Rao, P. and Moldave, K. (1967). *Biochem. biophys. Res. Commun.* **28**, 909–913.

Ravel, J. M., Wang, S. F., Heinemeyer, C. and Shive, W. (1965). *J. biol. Chem.* **240**, 432–438.

Reich, E. (1964). *Science, N.Y.* **143**, 684–689.

Robinson, E. and Brown, R. (1953). *Nature, Lond.* **171**, 313.

Rosset, R., Monier, R. and Jolien, J. (1964). *Bull. Soc. chim. biol.* **46**, 87–109.

Sager, R., Weinstein, I. B. and Ashkenazi, Y. (1963). *Science, N.Y.* **140**, 304–306.

Salas, M., Hille, M. B., Last, J. A., Wahba, A. J. and Ochoa, S. (1967). *Proc. natn. Acad. Sci. U.S.A.* **57**, 387–394.

Schwartz, J. H., Eisenstadt, J. M., Brawerman, G. and Zinder, N. D. (1965). *Proc. natn. Acad. Sci. U.S.A.* **53**, 195–200.

Schweet, R. S. and Allen, E. H. (1958). *J. biol. Chem.* **233**, 1104–1108.

Setterfield, G., Neelin, J. M., Neelin, E. M. and Bayley, S. T. (1960). *J. molec. Biol.* **2**, 416–424.

Siegel, M. R. and Sisler, H. D. (1965). *Biochim. biophys. Acta* **103**, 558–567.

Sissakian, N. M., Filippovich, I. I. and Svetailo, E. N. (1963). *Proc. Acad. Sci. USSR* (Eng. trans.) **147**, 1204–1205.

Spencer, D. and Wildman, S. G. (1964). *Biochemistry, N.Y.* **3**, 954–958.

Speyer, J. F., Lengyel, P., Basilio, C., Wahba, A. J., Gardner, R. S. and Ochoa, S. (1963). *Cold Spring Harb. Symp. quant. Biol.* **28**, 559–567.

Stephenson, M. L., Thimann, K. V. and Zamecnik, P. C. (1956). *Archs Biochem. Biophys.* **65**, 194–209.

Stocking, C. R. (1959). *Pl. Physiol., Lancaster* **34**, 56–61.

Stocking, C. R., Shumway, L. K., Weier, R. E. and Greenwood, D. (1968). *J. cell. Biol.* (in press).

Stout, E. R. and Mans, R. J. (1967). *Biochim. biophys. Acta* **134**, 327–336.

Stulberg, M. P. and Novelli, G. D. (1962). *In*, "Methods in Enzymology" (S. P. Colowick and N. O. Kaplan, eds). Vol. 5, pp. 703–707. Academic Press, New York and London.

Suarez, G. and Nathans, D. (1965). *Biochem. biophys. Res. Commun.* **18**, 743–750.

Sueoka, N. and Yamane, T. (1962). *Proc. natn. Acad. Sci. U.S.A.* **48**, 1454–1461.

Takanami, M. and Okamoto, T. (1960). *Biochim. biophys. Acta* **44**, 379–381.

Ts'o, P. O. P. (1962). *A. Rev. Pl. Physiol.* **13**, 45–80.

Ts'o, P. O. P., Bonner, J. and Vinograd, J. (1956). *J. biophys. biochem. Cytol.* **2**, 451–466.

Varner, J. E. and Chandra, G. R. (1964). *Proc. natn. Acad. Sci. U.S.A.* **52**, 100–106.

Wallace, J. M., Squires, R. F. and Ts'o, P. O. P. (1961). *Biochim. biophys. Acta* **49**, 130–140.

Warner, J., Rich, A. and Hall, C. (1962). *Science, N.Y.* **138**, 1399–1403.

Watts, R. L. and Mathias, A. P. (1967). *Biochim. biophys. Acta* **145**, 828–831.

Webster, G. C. (1957). *J. biol. Chem.* **229**, 535–546.

Webster, G. C. (1959a). *Archs Biochem. Biophys.* **82**, 125–134.

Webster, G. C. (1959b). *Archs Biochem. Biophys.* **85**, 159–170.

Webster, R. E., Engelhardt, D. L. and Zinder, N. D. (1966). *Proc. natn. Acad. Sci. U.S.A.* **55**, 155–161.

Weisblum, B., Benzer, S. and Holley, R. W. (1962). *Proc. natn. Acad. Sci. U.S.A.* **48**, 1449–1454.

Whaley, W. G., Mollenhaver, H. H. and Leech, H. (1960). *Am. J. Bot.* **47**, 401–449.

Williams, G. R. and Novelli, G. D. (1968). *Biochim. biophys. Acta* (in press).

Wood, W. B. and Berg, P. (1962). *Proc. natn. Acad. Sci. U.S.A.* **48**, 94–104.

Zachau, H. G., Dütting, D. and Feldman, R. (1966). *Angew. Chem., Intern. Ed. Engl.* **5**, 422.

CHAPTER 3

Polysomes: Analysis of Structure and Function

HANS NOLL

Department of Biological Sciences, Northwestern University,
Evanston, Illinois, U.S.A.

I. DISCOVERY AND GENERAL PROPERTIES OF POLYSOMES

Three different laboratories independently and simultaneously contributed the decisive observations which established the significance of the ribosomal aggregates and clearly spelled out the polysome concept. Experiments with reticulocytes showed that nascent protein which had been labelled in the intact cells was associated with a heterogeneous 170S ribosome fraction and not with 80S single ribosomes (Warner, Knopf and Rich, 1963). (This, incidentally, is a very sensitive test for judging whether breakdown has occurred during the preparation of polysomes.) Examining the material in the analytical ultracentrifuge, Gierer (1963) observed that what had appeared as a

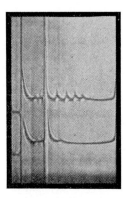

FIG. 1. Sedimentation in the analytical ultracentrifuge of reticulocyte polysomes before (top) and after (bottom) treatment with RNase. (From A. Gierer, *J. molec. Biol.* **6**, 148 (1963), with permission.)

broad 170S band in the sucrose gradient was now resolved into 5 sharp peaks (Fig. 1). Similarly, Wettstein, Staehelin and Noll (1963), trying to improve the amino acid incorporation activity of rat liver ribosomes in a cell-free system noted that the activity increased with the proportion of heavy aggregates present. The most active preparations contained very little 80S ribosomes and consisted mostly of faster sedimenting material. When analyzed in a sucrose gradient by the conventional discontinuous sampling methods, the material appeared to be spread over a wide region corresponding to 150–400S, with an indication of several discrete peaks ahead of the 80S ribosomes. Since the continuous spectrum of rapidly sedimenting ribosomes could have resulted from a reversible interaction and aggregation of single ribosomes, it was imperative to improve the resolution to the point where it permitted a clear-cut decision whether or not the suspected additional peaks were real. This was accomplished by introducing continuous spectrophotometric monitoring of the contents of the gradient tube. The sedimentation patterns traced by the strip chart recorder in Fig. 2a, b immediately revealed 6 well-resolved peaks spaced at regularly decreasing intervals and merging with an unresolved region toward the bottom of the centrifuge tube. The sedimentation coefficients corresponding to these peaks could be approximated by the relation $(S/80)^{\frac{3}{2}} = n$, expected for aggregates consisting of n $80S$ ribosomes. The same was found to hold true for the peaks observed in the analytical ultracentrifuge (Fig. 1) with preparations of reticulocyte ribosomes (Gierer, 1963).

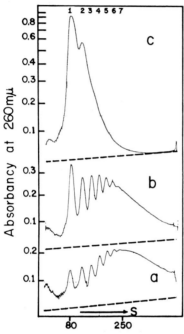

FIG. 2. Continuously recorded sedimentation patterns illustrating fragmentation of rat liver polysomes by traces of RNase. (a) C-ribosomes not treated with RNase, (b) incubated at 37°C for 5 min without RNase, (c) incubated at 37°C for 5 min in presence of $0 \cdot 012$ $\mu g/ml$ of RNase. (From F. O. Wettstein, T. Staehelin and H. Noll, *Nature, Lond.* **197**, 430 (1963), with permission.)

The relationship between aggregate structure and amino acid incorporation activity (Fig. 3) immediately suggested that messenger RNA was responsible for aggregate formation. The most attractive interpretation, namely that polysomes were single ribosomes strung together by a single continuous strand of messenger RNA, was strongly favored by the extreme sensitivity of the aggregates to traces of endonucleases and shearing forces. This explains most of the technical difficulties encountered in the early attempts to develop reproducible methods for the preparation of polysomes and, with many tissues, still presents a major problem. The effects of RNase and of shear (Oppenheim, Scheinbucks, Biava and Marcus, 1968) produced by

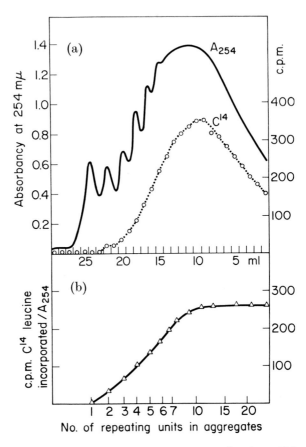

FIG. 3. Relationship between aggregate structure and amino acid incorporation activity. Fractions collected from the gradient were incubated at 37°C for 45 min with postmicrosomal supernatant, GTP and an ATP-generating system. The results plotted show in (a) total radioactivity incorporated as alkali-stable, acid-insoluble counts by each fraction, in (b) specific activity (cpm ^{14}C-leucine/0·1 mg RNA) as a function of aggregate size. (From H. Noll, T. Staehelin and F. O. Wettstein, *Nature, Lond.* **198**, 632 (1963), with permission.)

pipetting are illustrated in Figs. 2c and 4. Both treatments lead to a dramatic increase of single ribosomes at the expense of the larger aggregates.

Additional evidence that the structure linking ribosomes together into polysomes is in fact messenger RNA came from many different lines of evidence, some of which are summarized below.

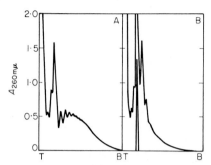

FIG. 4. The effect of shearing on polysome structure. A cytoplasmic extract from *Azotobacter vinelandii* was either centrifuged directly (A) or after pipetting 10 times with a small-bore 1-ml pipette (B). (From J. Oppenheim, J. Scheinbucks, C. Biava and L. Marcus, *Biochim. biophys. Acta* **161**, 386 (1968), with permission.)

A. Proportionality Between Polysome Size and Molecular Weight of mRNA

This was first demonstrated by Staehelin, Wettstein, Oura and Noll (1964) by determining the sedimentation coefficients of messenger RNA preparations derived from polysome classes of exactly defined size (Figs. 5–7). Since in polysomal mRNA accounts for only about 1%

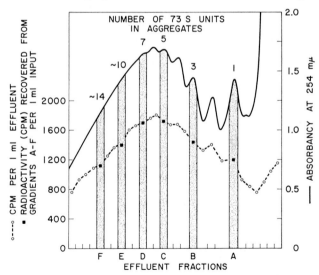

FIG. 5. Fractionation of polysomes according to aggregate size. (From T. Staehelin, F. O. Wettstein, H. Oura and H. Noll, *Nature, Lond.* **201**, 264 (1964), with permission.)

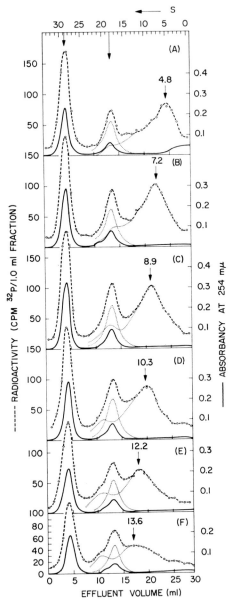

Fig. 6. Sedimentation analysis of messenger RNA corresponding to polysomes of different aggregate size. Fractions 1, 3, 5, 7, 10 and 14 from gradient shown in Fig. 5 were treated with sodium dodecyl sulfate and analyzed by zone velocity centrifugation (A–F). Number above arrow indicates sedimentation coefficient of mRNA peak. Dotted lines show interaction between 18S rRNA and mRNA. (From T. Staehelin, F. O. Wettstein, H. Oura and H. Noll, *Nature, Lond.* **201**, 264 (1964), with permission.)

of the total RNA (nearly all of which is ribosomal), optical methods are not sufficiently sensitive or specific to differentiate messenger from the bulk of ribosomal RNA. For this reason, the successful demonstration of the proportionality between mRNA and polysome size required methods that permitted the preferential labelling of mRNA. In most cases such conditions can be worked out as shown below.

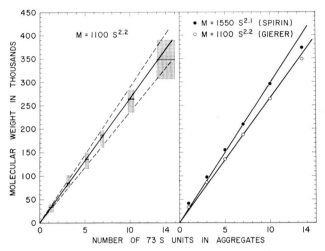

FIG. 7. Proportionality between molecular weight of messenger RNA and polysome size. Left: The shaded areas give an estimate of the maximal experimental errors, composed of the errors with respect to average aggregate size (Fig. 5) and average sedimentation coefficient (Fig. 6). The molecular weights corresponding to the horizontal bars were calculated according to Gierer from the sedimentation coefficients shown by arrows in Fig. 6. Right: The same data plotted according to the formulas of Gierer and Spirin. (From T. Staehelin, F. O. Wettstein, H. Oura and H. Noll, *Nature, Lond.* **201**, 264 (1964), with permission.)

B. Reduction in Average Messenger Size upon Fragmentation of Polysomes with Traces of RNase

Messenger RNA *in polysomes* can be labelled selectively by relatively short exposure to radioactive precursors (pulse labelling). This effect may result either from a more rapid metabolic turnover of mRNA compared to the other polysome-associated RNAs, or from a delayed entry into polysomes of rRNA relative to mRNA, or from a combination of both effects. Even though in rapidly growing bacteria, the metabolic instability of messenger RNA has been regarded as the main factor responsible for selective labelling, recent experiments (Mangiarotti and Schlessinger, 1967) show that the delayed entry of radioactive ribosomes into polysomes is, at least initially, more important. Since the assembly

of ribosomes from rRNA and ribosomal proteins is a relatively slow
process, there will be a lag between addition of label and the time at
which the newly assembled ribosomes containing fully labelled rRNA

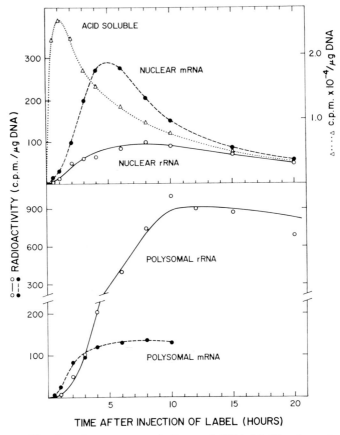

FIG. 8. Kinetics of messenger and ribosomal RNA labelling in nucleus and
cytoplasm. The specific radioactivity of the nuclear and polysomal fractions were
normalized with respect to DNA, assuming that the mass ratio of total ribosomal
RNA/total DNA = 3. (From H. Noll, Polysome Organization as a Control Element.
In, "Developmental and Metabolic Control Mechanisms and Neoplasia", 19th
Annual Symposium on Fundamental Cancer Research published for the University
of Texas M. D. Anderson Hospital and Tumor Institute, Houston, Texas, p. 67, The
Williams & Wilkins Co., Baltimore (1965), with permission.)

molecules begin to appear in polysomes. During this lag most of the
radioactivity appearing in polysomes is in newly synthesized mRNA
molecules. In the case of cells from non-growing mammalian tissues
(e.g. rat liver), the delayed entry of rRNA into polysomes has long been
shown to be the basis for the selective labelling of mRNA (Noll, 1965).

This is illustrated in Fig. 8, which shows that the labelled mRNA appears at maximal rate after only about an hour, whereas it takes nearly 6 h for the radioactivity of newly synthesized rRNA to reach its maximal rate of rise. As expected, this delay is not apparent in the

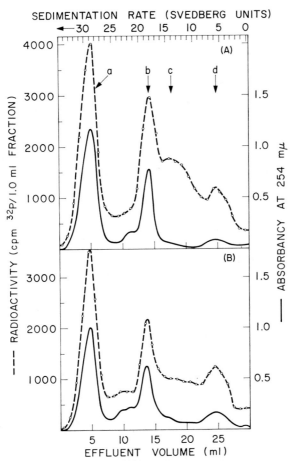

FIG. 9. Sedimentation pattern of pulse-labelled RNA components prepared from (A) normal and (B) RNase-treated rat liver polysomes. (From T. Staehelin, F. O. Wettstein, H. Oura and H. Noll, *Nature, Lond.* **201**, 264 (1964), with permission.)

nucleus where the measurements do not discriminate between radio-activity in free and ribosome-associated rRNA. The rapidly labelled or early labelled RNA associated with polysomes is characterized by a very high specific activity and heterogeneous size. The size distribution covers a range of 8–20*S* (peak c in Fig. 9A) and corresponds to the molecular weights of 10^5 to 6×10^5 expected for messengers coding for single polypeptide chains. The messenger peak disappears if, prior to disruption with sodium dodecyl sulfate (SDS), the polysomes are fragmented into smaller aggregates and single ribosomes by exposure to traces of RNase (Fig. 9b). Moreover, the messenger peak is absent or greatly reduced with respect to radioactivity and sedimentation velocity if, for SDS treatment, monomers are used that have been generated from polysomes during *in vitro* protein synthesis and, hence, are free of mRNA (Staehelin *et al.*, 1964).

C. In Vivo Breakdown of Polysomes Following Inhibition of Messenger RNA Synthesis with Actinomycin

If synthesis of messenger RNA is blocked by actinomycin, new polysomes cannot be formed and the existing polysomes break down at a rate reflecting the survival rate of mRNA synthesized prior to actinomycin treatment (Staehelin, Wettstein and Noll, 1963).

D. Formation of Polysomes from Ribosomes and Virus RNA

Reformation of polysomes in actinomycin-treated HeLa cells is observed upon infection of the cells with poliovirus (Fig. 10). These polysomes are engaged in the synthesis of poliovirus specific protein (Penman, Scherrer, Becker and Darnell, 1963). The *in vitro* formation of polysomes has also been accomplished with RNA from f2 bacteriophage (Capecchi, 1966; Engelhardt, Webster and Zinder, 1967).

E. Function of Polysomes: Tape Mechanism of Translation

While these findings established the structural concept that polysomes consist of single ribosomes held together by messenger RNA, they did not disclose much information about the way they function in the assembly of a polypeptide chain. The question as to the nature of the process by which amino acids are assembled on a template was asked in its most general form by Dintzis (1961) long before the discovery of polysomes. In a now classical paper, Dintzis proved that translation was initiated at the amino terminal of a polypeptide chain and advanced to its carboxyl end by the stepwise addition of amino acids. Dintzis' findings had far-reaching implications. The demonstration that the synthesis of a polypeptide chain proceeds from a fixed starting point at or near its amino terminus suggested that chain initiation was an important event, presumably triggered by a special signal, and in one stroke explained the tantalizing puzzle of how a non-overlapping code could be read unambiguously with correct phasing. Although Dintzis' results questioned the necessity, if not the validity, of the ideas visualizing the template as a static structure rigidly fixed and wrapped around the ribosomal surface, this notion continued to dominate the thinking for another two years (Stent, 1963). The decisive turn which

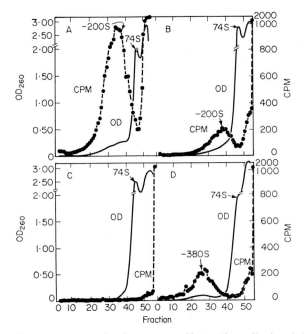

FIG. 10. The appearance of polysomes specific to the poliovirus-infected cell. HeLa cells suspended in a medium containing actinomycin were infected with poliovirus (~ 50 plaque-forming units per cell) 3 h after addition of actinomycin and analyzed for polysomes at 2 (B), 2·75 (C) and 3·75 (D) h after infection. Non-infected cells exposed to actinomycin for 5 h (A) served as controls. Protein synthesis was followed by the incorporation of radioactive amino acids added at the time of infection. (From S. Penman, K. Scherrer, Y. Becker and J. E. Darnell, *Proc. natn. Acad. Sci. U.S.A.* **49**, 654 (1963), with permission.)

prepared the ground for the tape mechanism of translation was Gilbert's recognition (already anticipated by Dintzis) that each ribosome was the center of growth of one and only one polypeptide chain (Gilbert, 1963). At this site the transfer RNA molecules were thought to deliver the amino acid called for by the messenger to the growing point of the chain. Once the idea of a fixed growing point on the ribosome is accepted, the postulate of a translational movement of the messenger tape relative to the growing point follows of necessity. The sequential, stepwise addition of amino acids established by Dintzis also becomes a necessary consequence of this model.

Direct evidence in favor of the tape model was published shortly after the first papers describing the discovery of polysomes. In a series of kinetic experiments designed to test the predictions of the model in a rigorous fashion, Noll, Staehelin and Wettstein (1963) verified the predictions of the model in all essential aspects. On the basis of their results, they also advanced the proposal that the hitherto puzzling function of GTP was to serve as cofactor and energy source for advancing the messenger tape. The experimental test of the tape model exploited the observation that most cell-free polysomal incorporation systems fail to support chain initiation and, thus, are limited to chain extension and completion. Evidently, in the absence of reattachment of ribosomes to the beginning of the message, polysomes must break down progressively by the successive release of ribosomes that have completed their read-out cycle. The model makes precise predictions about the distribution of the nascent protein label in relation to the polysome pattern. The most important of these predictions are: (i) the release of single ribosomes from polysomes should be proportional to the number of amino acids incorporated per ribosome in a polysome; (ii) the specific activity in the soluble pool (= radioactivity released into soluble pool of finished chains/ribosome released from polysome) should rise at exactly half the rate of the specific activity of the ribosome-bound protein and (iii) the fraction of radioactive nascent protein on ribosomes should change during chain completion in a characteristic manner which is determined by the initial size distribution of polysomes (Noll, 1965). All of these predictions have been verified by quantitative analytical measurements (Noll and Staehelin, 1969). For a detailed derivation and discussion of these relationships, the reader is referred to the article by Noll (1969) in "Protein Biosynthesis".

II. ANALYSIS OF POLYSOMES BY ZONE VELOCITY SEDIMENTATION IN SUCROSE GRADIENTS

So far sedimentation techniques have been the only means for detecting and assaying polysomes for biochemical studies. Moreover, as evident from the preceding section, most of the conclusions concerning the structure and function of polysomes have been derived from the analysis of sedimentation data. In view of the prominent place and generally low standards of practice of ultracentrifugation among the techniques in protein biosynthesis, a sufficiently detailed discussion of this subject seems desirable. Although polysomes have been studied with the analytical ultracentrifuge, equipped with either Schlieren or

UV-optics, zone velocity centrifugation in sucrose gradients has been used much more extensively because it combines high resolution with the advantages of a preparative instrument. Indeed, owing to recent improvements, higher resolution is now attainable by this method than by the conventional techniques of analytical ultracentrifugation. Polysomes may be regarded as an ideal test object to gauge the performance and resolution of analytical centrifugation methods since

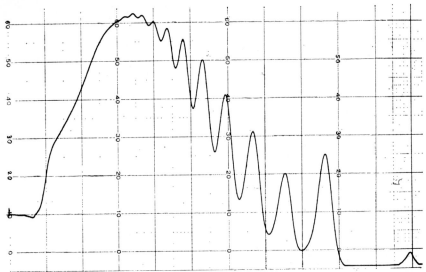

FIG. 11. Sedimentation pattern of mouse liver polysomes in 4·8 ml Spinco rotor SW-65. Centrifugation was for 15 min at 65,000 rev/min and 8°C in a non-isokinetic sucrose gradient (concave exponential, $C_R = 0$, $C_m = 1·3M$). (From H. Noll, *Analyt. Biochem.* **27**, 130 (1969), with permission.)

they represent a mixture of macromolecules covering a range of molecular weights which increase in discrete steps according to a regular series. Hence, as we have already seen, zonal sedimentation produces a progression of bands, spaced at regularly decreasing intervals, in accordance with the growing number of ribosomes on messenger RNA. The resolving power of the method, therefore, can be expressed by the number of bands that are resolvable. From published data, it appears that the analytical method is capable of resolving, at most, 5 bands, whereas zone velocity centrifugation with proper equipment succeeds in resolving up to 11 bands (Fig. 11).

Unfortunately, the practice of the method has rarely exploited its analytical power, either with respect to resolution or the quantititave determination of sedimentation rates. The two examples shown in Fig. 12 are fairly typical for so-called polysome sedimentation patterns

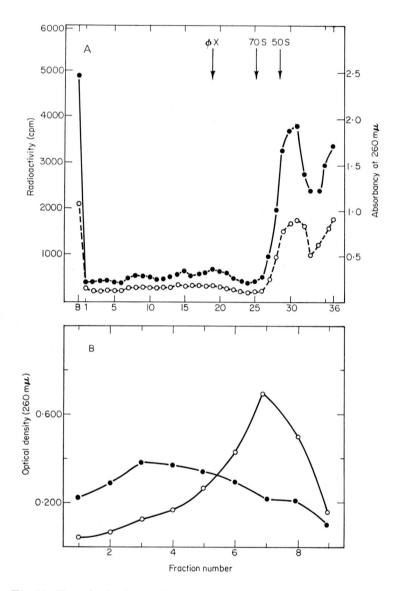

FIG. 12. Unresolved sedimentation patterns frequently attributed to polysomes in published work. A. Sedimentation diagram of extracts from *Escherichia coli* cells that had been labelled with ^{14}C-uracil for two generations. The upper curve shows the acid-insoluble radioactivity (solid circles), the lower curve the absorbancy at 260 mμ (open circles). B. Sedimentation pattern of liver ribosomes. The open circles shows the distribution after starvation, the solid circles after starvation and re-feeding.

most frequently encountered in the literature: an indication of a monomer peak followed by an amorphous broad band of UV-absorbing (or light scattering) material. In the absence of the progression of peaks characteristic of polysomes, identification of the material corresponding to the broad band as polysomes with any reasonable degree of certainty is not possible, and the unresolved patterns could as well be attributed to any other polydisperse material that happens to absorb or scatter light at 260 mμ, as for example glycogen or unspecific aggregates of ribosomes. Not even the demonstration of an UV-absorbancy shift toward the monomer region after treatment with RNase is acceptable proof for the polysome nature of the material in question, unless the monomer peak is sufficiently well resolved to allow its unambiguous identification and to eliminate the possibility that the absorbancy is due to fragments of ribosomal RNA produced by RNase.

A. General Principles

Zone velocity centrifugation fractionates macromolecules according to size, density and shape. If solute molecules are exposed to a centrifugal field, the force acting on them is opposed by the forces of friction and buoyancy. Since with increasing molecular weight the surface/mass ratio and hence the relative frictional resistance diminishes, the sedimentation rate will increase with molecular weight within a family of molecules of similar shape and density. Likewise, sedimentation or flotation rates increase proportionally to the difference in density between particle and surrounding medium. Molecules that have the same density as the medium will not sediment and those of lower density will float.

If a solution containing a mixture of molecules to be separated is layered in a narrow zone over a liquid column and spun at high speed in a swinging bucket rotor, the solute molecules will move outward and separate into bands according to their individual sedimentation rates. Important conditions are that the liquid column be free of convection and that the sample layer be of a lower density than the top of the liquid column, while the molecules to be analyzed, must of course, be denser than the liquid column. Stabilization of the liquid column against convection is achieved by introducing a density gradient which restricts the movement of the solvent molecules along the axis of the tube.

B. Resolution

An important factor determining resolution is the band width which in turn depends on the molecular weight and concentration of the macromolecules in the sample, the volume and overall density of the sample layer and the shape of the gradient. To maintain hydrostatic stability, it is necessary to prevent local density inversions which occur whenever the trailing edge of a band is allowed to reach a higher density than its leading edge. In a theoretical treatment of this subject, Berman (1966) has shown that a band that is initially stable can become unstable during migration. In general, the following rules apply. Increasing the slope of the gradient causes band sharpening, since the viscous drag rises enormously with increasing sucrose concentration. This band compression may reduce the band capacity (i.e. the maximum mass of particles that can be carried in a stable band of a given width) to such an extent that density inversion occurs. For this reason better resolutions are usually achieved by using relatively wide top zones and low sample concentrations rather than vice versa. Poor resolution is almost always the result of overloading. On the other hand, astonishingly sharp bands can be obtained with very large loads, if the volume of the sample layer is expanded and the slope of the gradient increased. Thus, it is possible to produce well-resolved polysome patterns by layering a volume of 1·5 ml containing 10 mg of material over a 3·5 ml gradient. If the top zones are wider than about 0·25 ml/cm², they should be stabilized against convection by means of a gradient prepared by gradually diluting a small volume of sucrose solution with the sample solution. The starting concentration of this sucrose solution must be equal to or slightly less than that of the top of the gradient and, as it is diluted with sample, it is carefully layered over the gradient. Thus, the sample is introduced in the form of an inverted gradient. (However, for stability the *total* density of the top zone must still *decrease* towards the top.) This measure also protects against density inversion which would be inevitable if all the particles of a large top zone were allowed to accumulate near the viscosity barrier at the interface between gradient and top zone.

   Although there have been fancy theoretical treatments involving the use of mass transport equations to predict the influence of loading on band width and resolution, they have so far been of little value to the experimenter, as these mathematical formulations have not been reduced to numerical values nor have they been tested experimentally. A rough empirical test may be used for a quick determination whether the solution to be layered over the gradient will form a stable zone. A few ml of sucrose of a concentration corresponding to that of the top layer is placed in a test tube and a drop of the test solution is allowed to fall into it from about 1 cm above the surface. It should bounce back and float; if it sinks, it will be too dense for the top layer of the gradient. McConkey (1967) has given, without derivation, the following formula to estimate the band capacity:

$$Y = \frac{dc}{dh} Ah^2/2,$$

in which Y is the total amount of sample in mg that can be applied, h the thickness of the sample layer in mm, dc/dh the concentration gradient of the sample in $mg/cm^3/mm$, and A the cross-sectional area in cm^2. As pointed out by McConkey, this formula should only be taken as a rough estimate. In fact, the validity of this formula has apparently never been verified by systematic experimental tests. Although it would be most desirable to know with reasonable accuracy the maximal capacity compatible with optimal band separation under a given set of conditions, unfortunately there are not sufficient data available on this point.

Another limitation of resolution is imposed by the fact that with the most commonly used gradients and rotors the molecules slow down as they move away from the top layer, for the viscous drag of the medium increases more rapidly than the centrifugal force. As a result, the particles tend to pile up toward the bottom of the centrifuge tube. Consequently, in most gradients the separation between two components fails to improve with sedimentation beyond a third to a half of the entire length of the tube and the potential resolution offered by the available path length remains largely unexploited. To remedy this situation and at the same time facilitate the determination of sedimentation coefficients, Noll introduced isokinetic sucrose gradients which were calculated to keep the sedimentation rate of particles of a given density constant throughout the length of the tube (Noll, 1967).

C. Determination of Sedimentation Coefficients

The determination of sedimentation coefficients in sucrose gradients has been largely a matter of guesswork. Very often grossly erroneous values were derived by linear inter- or extrapolation relative to an internal standard. In order to appreciate the salient points of this problem, a short discussion of some of the quantitative aspects is necessary. At a given temperature, the rate at which a particle of density D_p sediments through a medium of density D_m and viscosity η_m in a centrifugal field $\omega^2 r$ is

$$dr/dt = s_{20,w}(\omega^2/A)(D_p - D_m)r/\eta_m. \tag{1}$$

The term $A = (D_p - D_{20,w})/\eta_{20,w}$ has been introduced to normalize measurements with respect to temperature and viscosity of water at 20°C; it is constant for any given D_p. Although the sedimentation coefficient is an important and characteristic physical constant of a macromolecule, the information it provides is somewhat ambiguous as it lumps together contributions from a number of different and often divergent properties such as molecular weight, density and shape. However, if any two of these parameters are kept constant within a series, the change in sedimentation coefficient will reflect the change in one particular property only and thus disclose very precise information about the molecules under study. As we shall see, this is often the case.

In order to determine the sedimentation coefficient of a particle from the distance travelled in a given time, we have to integrate Eq. (1), unless the sedimentation velocity is constant as in the isokinetic gradients. For molecules sedimenting in an aqueous solution, the integration is simple, since the force acting on the molecules is proportional to their distance from the rotor center. However, when Eq. (1) is applied to the sedimentation kinetics in a sucrose gradient, the integration becomes complicated, since D_m and η_m are no longer constant but functions of the changing sucrose concentration. To complicate matters further, the change in the medium might alter the shape and density of the sedimenting molecules, e.g. osmotic particles might shrink, etc.

In the case of arbitrary, non-isokinetic gradients, sedimentation coefficients are more easily determined by a simple empirical calibration method than by mathematical integration. A general method that gives fairly accurate values and is applicable to all particles of equal density and osmotic behavior and to any given rotor type and gradient is described below. Although this method requires a valid internal standard of known sedimentation coefficient, it is, unlike the mathematical methods, not dependent on the validity of whatever assumptions are made about the hydrodynamic behavior of the molecules under investigation.

After choosing a certain rotor type and gradient, we calibrate the gradient with a standard which must have the same density as the particles whose sedimentation coefficient we wish to determine. This is done by measuring the distance travelled by the standard as a function of centrifugation time, using sufficient different time intervals to cover the entire length of the tube. Several standards of different S-values but the same density may be used in the same tube, e.g. 30S, 50S and 70S ribosomes from E. $coli$. We then plot the distance travelled (d) or the effluent volume corresponding to the band center, against the product obtained by multiplying the known sedimentation constant (S) with the experimentally chosen centrifugation time (t) and the square of the angular velocity (ω^2).

$$d = St\omega^2. \tag{2}$$

The resulting calibration curve, reproduced in Fig. 13, is characteristic for a given gradient and rotor and, as already mentioned, for molecules

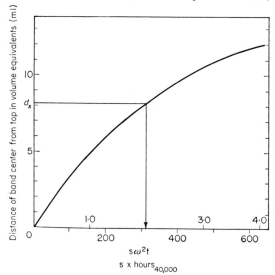

Fig. 13. Calibration of non-isokinetic sucrose gradients. The measurements were carried out with an IEC SB-283 rotor, using *E. coli* ribosomes as calibration standard. The curve describes sedimentation in a convex exponential gradient prepared with the device shown in Fig. 14. A total volume of 12 ml 1·4M sucrose was added to the mixing chamber containing 10 ml of 0·3M sucrose. The upper scale of the abscissa is calibrated in units of cm × sec for S, $2\pi \times$ r.p.s. (radians per sec) for ω and sec for t. The lower scale has been calibrated in units of 10^{-13} (Svedberg) at a rotor speed of 40,000 rev/min. Determination of the S-value of an unknown ribosomal peak: according to the calibration curve, the distance of sedimentation d_x observed after 2 h of centrifugation at 30,000 rev/min corresponds to a value on the abscissa of 2·0 (upper scale). Hence

$$s\omega^2 = 2\cdot0 \quad \text{or} \quad \frac{2}{(2\pi \times 500)^2 \times 7200} = 280 \times 10^{-13}\,\text{cm sec} \quad \text{or} \quad 280S.$$

If we use the lower scale which is calibrated for a rotor speed of 40,000 rev/min, a correction for the lower rotor speed of 30,000 rev/min must be made. Thus, 2 h at 30,000 rev/min are equivalent to 1·12 h at 40,000 rev/min ($2 \times 30,000^2/40,000^2$) and hence $S \times 1\cdot12 = 315$ and $S = 315/1\cdot12 = 280$.

of the same density. It usually exhibits a convex curvature which, for a given gradient, becomes more convex with increasing particle density. The S-value of the unknown is obtained from the graph by finding the value $St\omega^2$ on the abscissa that corresponds to the experimentally determined distance of migration d_x. Since both the centrifugation time t and centrifugal force ω^2 are known, S can be immediately computed. S-values of polysome peaks determined by this method are in close agreement with the values obtained in the analytical ultracentrifuge (Table 1).

TABLE 1

Comparison of sedimentation coefficients of polysomes determined in the analytical ultracentrifuge and in sucrose gradients by the calibration method illustrated in Fig. 13

Polysome size $(80S)_n$	S-values of rat liver polysomes Analytical ultracentrifuge*	Calibrated sucrose gradient
$n = 1$	83	81
2	123	120
3	154	153
4	183	180
5	211	207

* Data from Pfuderer, Cammarano, Holladay and Novelli (1965).

D. Isokinetic Gradients

The difficulties of computing S values in an arbitrary sucrose gradient by mathematical integration disappear if we select an isokinetic gradient, i.e. a gradient in which all particles of a given density sediment at a constant rate. The equation for isokinetic sedimentation is easily derived by demanding that the second term in Eq. (1), $r(D_p - D_m)/\eta_m$, must remain constant with increasing r. If the initial conditions are defined by the distance r_t of the meniscus from the center of rotation, by the choice of the sucrose concentration C_t at the top of the gradient and the corresponding values of D_t, η_t, it follows from the constant velocity condition that at any distance r in the gradient

$$\eta_{m(r)}/\eta_t = r[D_p - D_{m(r)}]/r_t[D_p - D_t]. \qquad (3)$$

If we now insert into Eq. (3) the functions $\eta^*(C)$, $D^*(C)$ describing viscosity and density as functions of the sucrose concentration, we obtain the desired isokinetic gradient

$$\eta^*(C) = \frac{\eta_t}{r_t(D_p - D_t)} r[D_p - D^*(C)]. \qquad (4)$$

This equation is transcendental and cannot be solved explicitly for C. The isokinetic gradient described by the function $C = f(r)$ may be evaluated by computer or graphical methods, using empirical expressions for the viscosity and density functions. The shape of the gradient is convex and can be approximated by the type of exponential gradient generated by a simple constant volume mixing device consisting of mixing chamber and burette (Fig. 14). The parameters for making isokinetic gradients corresponding to several commercial swinging bucket rotors, top concentrations and temperatures have been determined by computer and are tabulated in Table 2.

Apart from the errors introduced in preparing the sucrose solutions,* the accuracy of isokinetic gradients depends largely on the equipment used to produce the gradients, the reproducibility of the volume in the gradient tubes and the uniformity of the tube geometry. Variations in the wall thickness of the plastic centrifuge tubes may cause considerable variations in the distance of the top layer from the center of rotation, especially with small tubes in the 3·5–5 ml range. An additional error is introduced by the deformation of the plastic tubes and, possibly, of the supporting titanium buckets, expected at the extremely high forces generated at angular velocities in the 40,000–60,000 rev/min range. Nevertheless, tests with polysomes in two different laboratories with different rotors of similar capacities (12–14 ml) showed that the expected constancy of sedimentation velocity held true over the entire tube length, provided the isokinetic gradient of the correct density was selected (McCarty, Stafford and Brown, 1968). The strip chart recordings of the sedimentation patterns in Fig. 15 show that the resolution is excellent and that with increasing time of centrifugation the separation between peaks continues to increase throughout the whole length of the tube, in contrast to the piling up of the components toward the bottom observed in conventional gradients. The constancy of sedimentation rate is documented in Fig. 16 by the proportionality between S-values and the distance travelled in a given time by the various polysome species and in Fig. 17 by the linearity of travel with centrifugation time. In addition, Fig. 17 illustrates the importance of selecting the isokinetic gradient corresponding to the correct particle density.

E. Preparation of Gradients

There are three basic types of gradients that represent simple mathematical functions and are easy to prepare with ordinary laboratory equipment, namely (a) linear, (b) convex exponential and (c) concave exponential gradients (Fig. 18). All three types of gradients

---

* We have found it advisable to verify the sucrose concentration of the stock solutions by weighing a known volume, since the water content of the crystalline preparations seems to vary.

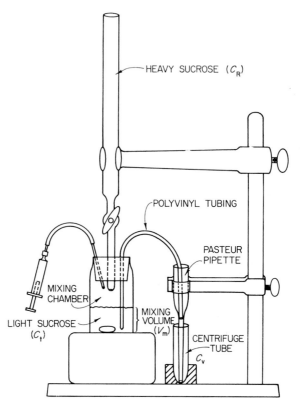

Fig. 14. Exponential gradient maker. The burette contains sucrose solution of concentration C_R and is connected through an air-tight rubber stopper to a mixing flask with magnetic stirring bar. A thin piece of polyvinyl tubing connects the mixing flask with the gradient tube. A small syringe is attached to the mixing chamber through a short piece of tubing to adjust in a reproducible way the amount of pressure at the beginning. The mixing flask has been filled with a sucrose solution of a concentration C_m equal to that desired at the top (or bottom) of the gradient. The procedure is started by increasing the air pressure in the mixing chamber with the syringe until the solution has risen to a fixed point in the ascending portion of the outlet tubing. The tubing between the syringe and mixing flask is then clamped off. The stopcock of the burette is now opened to add the heavy (or light) sucrose solution with rapid stirring. As the volume V added from the burette increases, a corresponding volume of the mixture is displaced into the gradient tube. The end of the outlet tubing from the mixing chamber is inserted into a capillary pipette whose tip must touch the bottom of the gradient tube so that the light sucrose introduced first is allowed to float to the top as solution of increasing density is fed in. At the end of the procedure the capillary pipette is carefully pulled out through the gradient. For the preparation of isokinetic gradients, the parameters V_m, C_R and C_t have to be chosen according to rotor dimensions and particle density (Table 2).

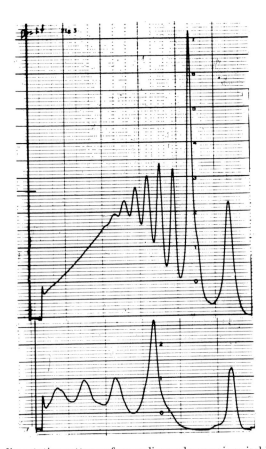

FIG. 15. Sedimentation patterns of mouse liver polysomes in an isokinetic gradient. Input: $4A_{260}$ units in 0·2 ml. Recorded after 1 h (top) and 2 h (bottom) centrifugation at 40,000 rev/min and 2°C. Gradient calculated for $D_p = 1·4$ and $C_t = 15\%$ sucrose. The strongest peak corresponds to the 80S monomer species. Centrifugations were carried out with the 6 × 14 ml swinging bucket rotor No. 283 in a B-60 ultracentrifuge of the International Equipment Company. The centrifuge was set to the desired top speed and decelerated with the automatic brake on. The gradient tubes were punctured and the absorbancy of the effluent at 260 mμ was scanned and recorded automatically with the IEC-Gilford high-resolution gradient analyzing system. Sedimentation rates were determined from the strip chart recordings by measuring the distance between the peak center and the top of the gradient. To indicate the top of the gradient, 0·2 ml of a 0·2M sucrose solution containing 0·5 mg/ml of RNA as an absorbancy marker was layered over the gradient immediately before collection (peak on extreme right). (From H. Noll, *Nature, Lond.* **215**, 360 (1967), with permission.)

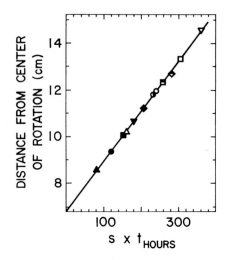

FIG. 16. Test for constancy of polysome sedimentation rates in an isokinetic gradient. Mobilities of polysome peaks determined from recordings in Fig. 15 are plotted against $S \times t$. Monomer to tetramer: $80S$ △ ▲, $119S$ ○ ●, $152S$ □ ■, $180S$ ▽ ▼ (solid symbols = 1 h, open symbols = 2 h); pentamer to octamer: $207S$ ◆, $230S$ ◖, $258S$ ◈, $281S$ ▬ (all 1 h). (From H. Noll, *Nature, Lond.* **215**, 360 (1967), with permission.)

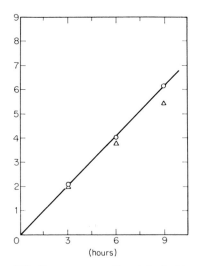

FIG. 17. Rat liver $80S$ ribosomes centrifuged in isokinetic gradients of (○) $1·41$ g/cm³ vs. (△) $1·51$ g/cm³. (From K. S. McCarty, D. Stafford and O. Brown, *Analyt. Biochem.* **24**, 314 (1968), with permission.)

3. POLYSOMES: ANALYSIS OF STRUCTURE AND FUNCTION   125

TABLE 2A

Parameters for isokinetic gradients in two IEC rotors

Particle density (g/cm³)	Rotor temp. (°C)	ROTOR IEC SB-405 4·0 ml gradient vol. 5·43–9·61 cm radius				ROTOR IEC SB-283 12·0 ml gradient vol. 6·74–14·3 cm radius			
		Conc. % w/v		Mixing chamber vol. (ml)	Reservoir conc. % w/v	Conc. % w/v		Mixing chamber vol. (ml)	Reservoir conc. % w/v
		Top	Bottom			Top	Bottom		
1·3	2	0	12·3	4·30	20·4	0	15·6	11·2	22·5
1·3	2	5	16·1	4·85	24·8	5	19·4	12·5	28·4
1·3	2	10	20·5	6·54	32·9	10	23·5	13·3	32·8
1·3	2	15	24·9	5·36	33·9	15	27·7	11·9	35·0
1·3	2	20	29·2	5·42	37·5	20	31·8	12·7	39·4
1·3	2	25	33·4	5·22	40·7	25	35·9	12·9	43·1
1·3	6	0	12·5	5·73	24·8	0	15·8	11·2	23·9
1·3	6	5	16·6	4·41	24·5	5	19·9	11·4	27·9
1·3	6	10	20·7	6·03	32·1	10	23·8	13·1	32·9
1·3	6	15	25·1	5·69	35·1	15	27·9	12·3	35·8
1·3	6	20	29·4	5·22	37·5	20	32·0	12·0	39·0
1·3	6	25	33·5	5·10	40·7	25	36·1	12·8	43·3
1·4	2	0	13·3	4·19	21·6	0	16·8	11·2	24·3
1·4	2	5	17·1	4·98	26·9	5	20·7	12·8	30·8
1·4	2	10	21·4	6·54	35·0	10	24·7	13·2	34·7
1·4	2	15	25·9	5·27	35·5	15	28·9	11·9	36·9
1·4	2	20	30·1	5·50	39·6	20	33·0	12·6	41·2
1·4	2	25	34·4	5·26	42·6	25	37·2	13·1	45·3
1·4	6	0	13·4	5·38	25·6	0	17·0	10·9	25·6
1·4	6	5	17·6	4·52	26·5	5	21·2	11·6	30·2
1·4	6	10	21·7	5·92	34·0	10	25·1	13·1	35·1
1·4	6	15	26·1	5·61	36·8	15	29·2	12·1	37·7
1·4	6	20	30·4	5·20	39·3	20	33·2	11·8	40·7
1·4	6	25	34·5	5·22	42·7	25	37·4	13·0	45·5

TABLE 2A (continued)

Unpublished computer data by D. Calhoun and H. Noll

Particle density (g/cm³)	Rotor temp. (°C)	ROTOR IEC SB-405 4·0 ml gradient vol. 5·43–9·61 cm radius				ROTOR IEC SB-283 12·0 ml gradient vol. 6·74–14·3 cm radius			
		Conc. % w/v		Mixing chamber vol. (ml)	Reservoir conc. % w/v	Conc. % w/v		Mixing chamber vol. (ml)	Reservoir conc. % w/v
		Top	Bottom			Top	Bottom		
1·5	2	0	13·9	4·12	22·3	0	17·7	10·3	25·6
1·5	2	5	17·7	5·08	28·3	5	21·5	12·9	32·3
1·5	2	10	22·0	6·43	36·0	10	25·5	12·9	35·7
1·5	2	15	26·5	5·19	36·3	15	29·7	11·9	38·2
1·5	2	20	30·7	5·49	40·7	20	33·7	12·4	42·0
1·5	2	25	35·0	5·40	44·0	25	37·9	12·9	46·2
1·5	6	0	14·1	5·15	26·1	0	17·9	11·9	26·8
1·5	6	5	18·3	4·57	27·8	5	22·1	11·7	31·6
1·5	6	10	22·3	5·93	35·1	10	25·9	12·9	36·3
1·5	6	15	26·7	5·53	37·7	15	30·0	11·9	38·7
1·5	6	20	30·9	5·14	40·2	20	33·9	11·6	41·6
1·5	6	25	35·1	5·35	44·2	25	38·1	12·9	46·6
1·6	2	0	14·3	4·06	22·9	0	18·2	10·4	26·6
1·6	2	5	18·2	5·14	29·3	5	22·0	12·9	33·2
1·6	2	10	22·5	6·35	36·7	10	25·9	12·8	36·3
1·6	2	15	26·8	5·12	36·8	15	30·2	11·9	38·9
1·6	2	20	31·1	5·46	41·3	20	34·1	12·2	42·6
1·6	2	25	35·3	5·44	44·8	25	38·3	12·7	46·8
1·6	6	0	14·5	4·99	26·3	0	18·5	10·9	27·7
1·6	6	5	18·7	4·59	28·6	5	22·6	11·7	32·6
1·6	6	10	22·8	5·91	35·9	10	26·4	12·8	37·0
1·6	6	15	27·1	5·43	38·3	15	30·5	11·8	39·3
1·6	6	20	31·3	5·09	40·7	20	34·4	11·5	42·2
1·6	6	25	35·5	5·40	45·0	25	38·5	12·9	47·3

TABLE 2A (continued)

Unpublished computer data by D. Calhoun and H. Noll		ROTOR IEC SB-405 4·0 ml gradient vol. 5·43–9·61 cm radius				ROTOR IEC SB-283 12·0 ml gradient vol. 6·74–14·3 cm radius			
Particle density (g/cm³)	Rotor temp. (°C)	Conc. % w/v		Mixing chamber vol. (ml)	Reservoir conc. % w/v	Conc. % w/v		Mixing chamber vol. (ml)	Reservoir conc. % w/v
		Top	Bottom			Top	Bottom		
1·7	2	0	14·7	4·02	23·2	0	18·7	10·4	27·3
1·7	2	5	18·5	5·17	30·0	5	22·4	12·9	33·9
1·7	2	10	22·8	6·28	37·1	10	26·4	12·7	36·7
1·7	2	15	27·1	5·06	37·2	15	30·5	11·8	39·4
1·7	2	20	31·3	5·43	41·7	20	34·5	12·0	42·9
1·7	2	25	35·6	5·45	45·3	25	38·6	12·6	47·1
1·7	6	0	14·9	4·88	26·5	0	18·9	10·9	28·3
1·7	6	5	19·1	4·61	29·3	5	23·1	11·8	33·3
1·7	6	10	23·1	5·89	36·5	10	26·8	12·7	37·5
1·7	6	15	27·4	5·35	38·6	15	30·9	11·7	39·7
1·7	6	20	31·6	5·05	41·1	20	34·7	11·5	42·8
1·7	6	25	35·7	5·41	45·5	25	38·8	12·8	47·7
1·8	2	0	14·9	3·98	23·5	0	19·0	10·4	27·8
1·8	2	5	18·7	5·19	30·6	5	22·7	12·9	34·4
1·8	2	10	23·0	6·23	37·4	10	26·6	12·6	37·0
1·8	2	15	27·3	5·01	37·4	15	30·8	11·8	39·7
1·8	2	20	31·5	5·40	42·0	20	34·7	12·0	43·3
1·8	2	25	35·8	5·44	45·7	25	38·8	12·5	47·3
1·8	6	0	15·1	4·79	26·7	0	19·3	10·9	28·8
1·8	6	5	19·3	4·61	29·7	5	23·4	11·8	33·8
1·8	6	10	23·3	5·86	36·9	10	27·1	12·6	37·8
1·8	6	15	27·6	5·29	38·8	15	31·1	11·6	40·0
1·8	6	20	31·7	5·01	41·4	20	34·9	11·5	43·2
1·8	6	25	35·9	5·42	45·9	25	39·0	12·7	47·9

TABLE 2B

Parameters for isokinetic gradients in 4 Beckman rotors

| Data from McCarty et al. (1968) | | | Rotor SW 39 or 50L Delivery vol. 5·0 ml 5·61–9·80 cm | | Rotor SW 25.1 Delivery vol. 30 ml 6·49–12·90 cm | | Rotor SW 25.2 Delivery vol. 58 ml 7·50–15·3 cm | | Rotor SW 25.3 Delivery vol. 16·5 ml 7·55–16·19 cm | |
Particle density (g/cm³)	Rotor temp. (°C)	Conc. % w/v top	Mixing chamber vol. (ml)	Reservoir conc. % w/v	Mixing chamber vol. (ml)	Reservoir conc. % w/v	Mixing chamber vol. (ml)	Reservoir conc. % w/v	Mixing chamber vol. (ml)	Reservoir conc. % w/v
1·33	5	5	6·07	23·7	30·2	24·8	56·5	25·0	14·5	25·5
1·33	5	10	6·30	27·1	31·4	28·1	58·5	28·2	14·9	28·6
1·33	5	15	6·50	30·4	32·2	31·2	60·2	31·5	15·3	31·8
1·33	20	5	6·00	24·8	29·7	26·0	55·5	26·2	14·2	26·6
1·33	20	10	6·17	28·0	30·7	29·1	56·6	30·4	14·6	29·7
1·33	20	15	6·37	31·4	31·6	32·3	59·1	32·5	15·1	32·9
1·41	5	5	6·00	24·6	29·9	25·8	55·9	26·0	14·3	26·5
1·41	5	10	6·23	28·0	31·0	29·0	57·9	29·2	14·8	29·6
1·41	5	15	6·47	31·4	31·9	32·3	59·6	32·5	15·2	32·8
1·41	20	5	5·90	25·8	29·4	27·0	54·9	27·3	14·0	27·8
1·41	20	10	6·10	29·0	30·3	30·1	55·9	31·2	14·5	30·8
1·41	20	15	6·30	32·4	31·4	33·3	58·6	33·6	15·0	34·0
1·51	5	5	5·93	25·4	28·1	25·7	55·2	26·8	14·1	27·3
1·51	5	10	6·17	28·7	30·6	29·8	57·2	30·0	14·6	30·4
1·51	5	15	6·37	32·0	31·6	33·0	59·0	33·2	15·1	33·6
1·51	20	5	5·80	26·6	28·9	27·9	54·1	28·2	13·8	28·7
1·51	20	10	6·00	29·8	29·9	31·0	54·6	32·3	14·3	31·7
1·51	20	15	6·23	33·1	31·0	34·2	57·9	34·4	14·8	34·8

TABLE 2B (continued)

Data from McCarty et al. (1968)			Rotor SW 39 or 50L Delivery vol. 5·0 ml 5·61–9·80 cm		Rotor SW 25.1 Delivery vol. 30 ml 6·49–12·90 cm		Rotor SW 25.2 Delivery vol. 58 ml 7·50–15·3 cm		Rotor SW 25.3 Delivery vol. 16·5 ml 7·55–16·19 cm	
Particle density (g/cm³)	Rotor temp. (°C)	Conc. % w/v top	Mixing chamber vol. (ml)	Reservoir conc. % w/v	Mixing chamber vol. (ml)	Reservoir conc. % w/v	Mixing chamber vol. (ml)	Reservoir conc. % w/v	Mixing chamber vol. (ml)	Reservoir conc. % w/v
1·77	5	5	5·77	26·3	28·9	27·7	53·8	27·9	13·8	28·4
1·77	5	10	6·03	29·6	29·9	30·8	55·9	31·0	14·3	31·4
1·77	5	15	6·23	32·9	30·9	33·9	57·7	34·1	14·8	34·5
1·77	20	5	5·67	27·7	28·2	29·0	52·7	29·3	13·5	29·8
1·77	20	10	5·87	30·8	29·2	32·1	54·4	32·4	14·0	32·8
1·77	20	15	6·10	33·1	30·3	35·0	56·7	35·4	14·5	35·9
1·81	5	5	5·77	26·4	28·7	27·7	53·7	28·0	13·7	28·5
1·81	5	10	6·00	29·7	29·9	30·9	55·8	31·1	14·3	31·5
1·81	5	15	6·23	33·0	30·8	34·0	57·6	34·2	14·7	34·6
1·81	20	5	5·63	27·8	28·1	29·1	52·5	29·4	13·5	30·0
1·81	20	10	5·83	30·9	29·1	32·2	52·2	29·6	13·9	32·0
1·81	20	15	6·07	34·2	30·2	35·3	56·5	35·5	14·5	36·0
1·89	5	5	5·73	26·6	28·6	27·9	53·4	28·2	13·7	28·7
1·89	5	10	5·97	29·8	29·7	31·0	55·5	31·2	14·2	31·7
1·89	5	15	6·20	33·1	30·7	34·1	57·3	34·4	14·6	34·8
1·89	20	5	5·60	28·0	27·9	29·3	54·2	32·6	13·4	30·1
1·89	20	10	5·83	31·1	28·9	32·3	54·2	32·6	13·9	33·1
1·89	20	15	6·07	34·3	30·1	35·5	56·3	35·7	14·4	36·1

can be produced by continuously adding a sucrose solution of concentration C_R from a reservoir to a mixing chamber containing a sucrose solution of concentration C_v, while at the same time displacing the mixture into the gradient tube (Fig. 19). The initial concentration in

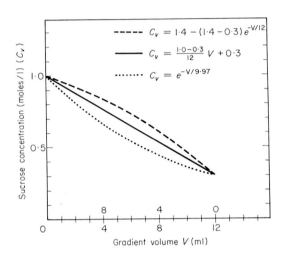

FIG. 18. Most frequently used types of gradients: linear ——————, convex exponential — — — — — and concave exponential .... The equations are represented by numerical examples for a 12-ml gradient tube. The scales on the abscissa are antiparallel because in the preparation of convex exponential gradients the lowest concentration enters the tube first (upper scale), while the highest concentration is added first in the case of concave exponential gradients.

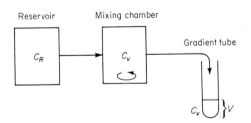

FIG. 19. Flow diagram for the preparation of linear and exponential gradients.

the mixing chamber $(C_v = C_i)$ changes as a function of the volume added and gradually approaches that in the reservoir $(C_v \to C_R)$.

The concentration of sucrose in the mixing chamber changes in a *linear fashion* as a function of the volume (dV) added from the reservoir if the volume of the mixture V_m contracts by the same amount dV_m

(Fig. 20). This process is described by the differential equation:

$$\frac{dC_v}{dV} = \frac{C_R - C_v}{V_m}. \tag{5}$$

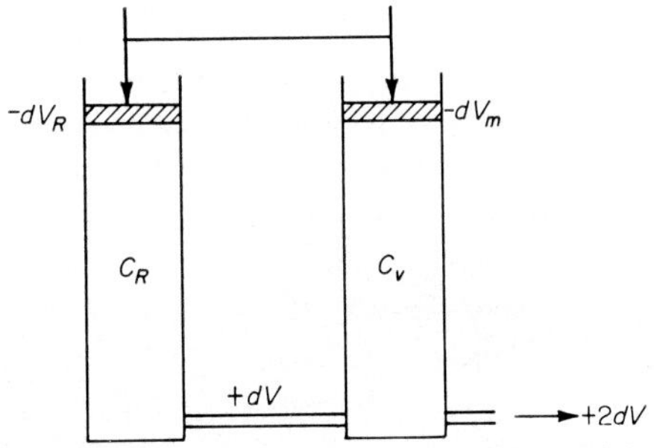

FIG. 20. Diagram illustrating the derivation of the differential equation describing the preparation of linear gradients.

Integrating and taking into account that $dV = -dV_m$

$$\ln |C_v - C_R| + K = \ln V_m. \tag{6}$$

The value of the integration constant is obtained from the initial conditions, $V_m = V_i$ and $C_v = C_i$:

$$K = \ln \frac{V_i}{|C_i - C_R|}$$

and inserted into Eq. (6):

$$\ln |C_v - C_R| + \ln \frac{V_i}{|C_i - C_R|} = \ln V_m.$$

Solving for C_v:

$$C_v = C_R + (C_i - C_R)(V_m/V_i). \tag{7}$$

Equation (7) expresses the concentration in the gradient as a function of the volume remaining in the mixing chamber. It is more convenient, however, to express it as a function of the volume V accumulating in the gradient tube. Thus, since $V_m = V_i - V/2$, we rewrite Eq. (7):

$$C_v = C_R + (C_i - C_R)(1 - V/2V_i),$$

$$C_v = \frac{C_R - C_i}{2V_i} V + C_i.* \tag{8}$$

---

* This equation applies strictly only if the volumes in the two vessels in Fig. 20 change at the same rate as, for example, in gradient machines consisting of closed vessels in which the liquid is displaced by means of two pistons. If two open communicating cylinders are used, $V_R \neq V_m$ at equilibrium because of the differences in hydrostatic pressure of sucrose solutions of different density. Equation (8) will then take the form:

$$C = \frac{C_R - C_i}{V_g} V + C_i \tag{8a}$$

in which V_g = final gradient volume.

6

The initial concentration in the mixing chamber, C_i, may be either larger or smaller than C_R, the concentration of the solution added from the reservoir. If $C_R > C_i$, then $C_i = C_t$ = concentration at top of gradient, whereas if $C_R < C_i$, $C_i = C_b$ = concentration at bottom of gradient.

If, in contrast to the case of linear gradients treated above, the mixing volume is kept constant during the procedure, an exponential gradient is obtained because as the concentration in the mixing chamber rises (or falls) each successive addition of the more (of less) concentrated sucrose solution causes a smaller amount of change in the mixture.

The general form of the differential equation describing exponential gradients is identical to Eq. (5) given above for linear gradients. The integrated form differs, however, because the mixing volume V_m remains constant:

$$\ln |C_v - C_R| + K = - V/V_m. \tag{9}$$

Equation (9) describes the concentration of the solution in the mixing chamber as a function of the volume V added from the reservoir or, since V_m = constant, as a function of the volume accumulating in the gradient. Hence, the initial conditions $V = 0$ and $C_v = C_i$, and the value of the integration constant $K = - \ln |C_i - C_R|$, inserted into Eq. (9) give:

$$\ln \left| \frac{C_v - C_R}{C_i - C_R} \right| = - \frac{V}{V_m} \quad \therefore \quad (C_v - C_R)/(C_i - C_R) = e^{-V/V_m},$$

$$C_v = C_R + (C_i - C_R) e^{-V/V_m}. \tag{10}$$

I. Linear Gradients

The simplest device for producing linear gradients consists of two cylindrical vessels connected by a short piece of tubing at the bottom (Fig. 21) as described by Britten and Roberts (1960). The connection between the two vessels can be opened or closed by means of a clamp or stop cock. One of the two vessels serves as a reservoir, the other is a mixing chamber fitted with a magnetic stirring bar and an outlet. The two vessels are filled with sucrose solutions of the balanced volumes V_t and V_b corresponding to the desired top and bottom concentrations of the gradient. The gradient is formed by drawing fluid from the outlet of the mixing chamber. There are two methods for carrying out this procedure, depending on whether the heavy or the light sucrose is put into the mixing chamber.

(1) Heavy sucrose in mixing chamber, light sucrose in reservoir: in this arrangement the concentration of the sucrose solution drawn from

the mixing chamber decreases at a linear rate. To prevent undesirable turbulence, it is necessary to put the tip of the outlet in contact with the inner wall of the gradient tube near the top so that the fluid is allowed to flow down the side as the tube is filled. This method is not feasible with water-repellent polyallomer tubes because their surface is not wettable.

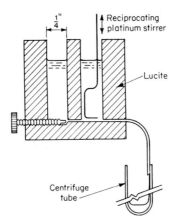

FIG. 21. Linear gradient maker. (From R. J. Britten and R. B. Roberts, *Science*, *N.Y.* **131**, 32 (1960), with permission.)

(2) Heavy sucrose in reservoir, light sucrose in mixing chamber: in this method the tip of the outlet, consisting of a capillary pipette, must touch the bottom of the gradient tube near the center so that the light sucrose introduced first is allowed to float to the top as fluid of increasing density is fed in. At the end of the procedure, the capillary pipette is carefully pulled out through the gradient. This method is suited for tubes made of water-repellent plastic.

The rate of flow is controlled either by a thumb screw or by the diameter of the outlet opening and should not be too rapid, otherwise it would cause turbulence. In the second procedure flow rates of up to 2 ml per minute are possible, whereas in the first method slower rates, about 1 ml per minute, are mandatory to prevent the stream flowing down the side of the tube from stirring up the gradient as it hits the surface. This tendency is reduced by tilting the gradient tube.

The hydrostatic pressure of the two fluid columns must be equal so that when the connection is opened there will be no flow in either direction. Balanced volumes (V_t, V_b) corresponding to the top and bottom concentrations are calculated from their densities (d_t, d_b):

$$V_t d_t = V_b d_b. \tag{11}$$

The densities are taken from Fig. 22 in which the density is plotted as a function of the sucrose concentration. The desired gradient volume V_g combines the two unknown volumes of the solutions in the reservoir and mixing chamber:

$$V_g = V_t + V_b. \tag{12}$$

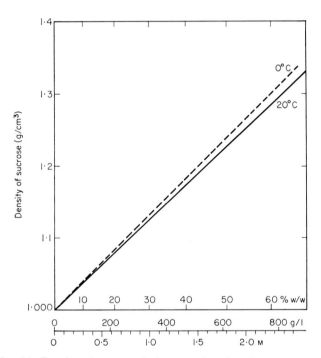

FIG. 22. Density of sucrose solutions as a function of concentration.

Combining Eqs. (11) and (12) and solving for V_b:

$$V_b = \frac{V_g d_t}{d_b + d_t}. \tag{13}$$

Since V_g is given, we immediately obtain V_t from Eq. (12).

If gradients are made by letting the gradient maker empty itself, a small portion will usually remain in the mixing chamber and reservoir, depending on the particular construction of the gradient maker. To calibrate the device for correct delivery volumes, the volume V_r of this residue must be determined and included in the total volume to be added to the gradient maker:

$$V_g' = V_g + V_r. \tag{14}$$

Evidently, the correction depends on whether the light or the heavy sucrose is delivered last. If the desired top and bottom concentrations of the gradient are C_t and C_b, the corrected concentrations (C_t', C_b') and volumes (V_t', V_b') to be added to the gradient maker are computed as follows:

(a) Light sucrose in mixing chamber:

$$C_t' = C_t; \quad C_b' = C_t + (V_g'/V_g)(C_b - C_t);$$

$$V_b' = (V_g' d_t)/(d_b' + d_t); \quad V_t' = V_g' - V_b'.$$

(b) Heavy sucrose in mixing chamber:

$$C_t' = C_b - (V_g'/V_t)(C_b - C_t); \quad C_b' = C_b;$$

$$V_b' = (V_g' d_t)/(d_b + d_t'); \quad V_t' = V_g' = V_b'.$$

Example: We wish to prepare 28 ml of a 0·3 to 1·0 M gradient. The gradient maker has a retention volume of 3·2 ml.

(a) Light sucrose in mixing chamber (introduction to bottom of tube):

$$\mathbf{C_t'} = C_t = \mathbf{0{\cdot}3}\text{M}; \quad \mathbf{C_b'} = 0{\cdot}3 + (31{\cdot}2/28)(1{\cdot}0 - 0{\cdot}3) = \mathbf{1{\cdot}08}\text{M};$$

$$\mathbf{V_b'} = (31{\cdot}2 \times 1{\cdot}04)/(1{\cdot}14 + 1{\cdot}04) = \mathbf{14{\cdot}8} \text{ ml.};$$

$$\mathbf{V_t} = 31{\cdot}2 - 14{\cdot}8 = \mathbf{16{\cdot}4} \text{ ml.}$$

(b) Heavy sucrose in mixing chamber (introduction to top of tube):

$$\mathbf{C_b'} = C_b = \mathbf{1{\cdot}0}\text{M}; \quad \mathbf{C_t'} = 1{\cdot}0 - (31{\cdot}2/28)(1{\cdot}0 - 0{\cdot}3) = \mathbf{0{\cdot}22}\text{M}$$

$$\mathbf{V_b'} = (31{\cdot}2 \times 1{\cdot}04)/(1{\cdot}13 + 1{\cdot}03) = \mathbf{14{\cdot}9} \text{ ml.}; \quad \mathbf{V_t} - 31{\cdot}2 - 14{\cdot}9 = \mathbf{16{\cdot}3} \text{ ml.}$$

A variety of linear gradient makers are available commercially. Some make use of the principle of a double-barrelled syringe pump instead of the hydrostatic pressure. Perfectly adequate linear gradient makers of the type shown in Fig. 21 can be constructed easily by glueing the barrels of disposable plastic syringes to a support plate. An important condition, not always observed in commercial products, is that the connection between the two vessels be slightly raised from the bottom, to prevent the fluid entering from the reservoir to tunnel through the mixing chamber without adequate mixing.

2. Exponential Gradients

Concave or convex exponential gradients are formed depending on whether the solution introduced into a mixing chamber of constant volume has a lower or higher concentration than the solution initially present in the mixing chamber. Thus, the same device can be used for

making both type of gradients. A simple apparatus consists of a
burette connected through an air-tight rubber stopper to a mixing
flask fitted with magnetic stirring bar and outlet tube through which
the gradient solution is displaced into the centrifuge tube as a
corresponding fluid volume is added to the mixing flask (Fig. 14).

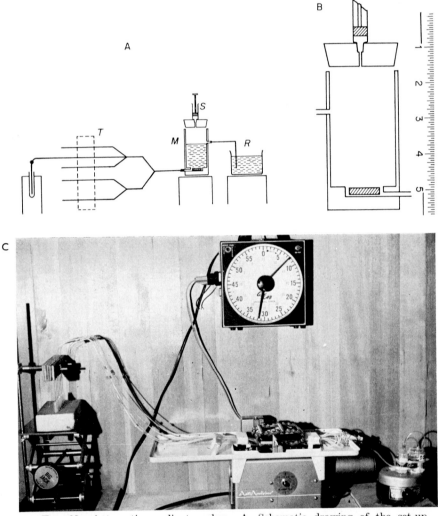

Fig. 23. Automatic gradient maker. A. Schematic drawing of the set-up
recommended by McCarty *et al.* B. Enlarged view of mixing chamber M.
C. Photograph of actual components assembled in the author's laboratory. (From
K. S. McCarty, D. Stafford and O. Brown, *Analyt. Biochem.* **24**, 314 (1968), with
permission, and H. Noll, unpublished information.)

McCarty *et al.* (1968) described a modification of this device which automatically produces an entire set of simultaneous and identical gradients. The apparatus, shown in Fig. 23, makes use of a multi-channel Technicon Auto Analyzer Pump which withdraws a desired volume from the mixing chamber and distributes it to a set of parallel tubes by means of a manifold. The use of a precision pump pulling the fluid out of the mixing chamber ensures greatest uniformity and reproducibility of the gradients. This is important in high resolution work, especially in conjunction with isokinetic gradients. The more primitive, burette-operated device shown in Fig. 14 appears to give satisfactory reproducibility when used for the preparation of gradients in the 3–15 ml range (Noll, 1967). However, as pointed out by Henderson (1969), if high precision is required, such as in the preparation of isokinetic gradients, certain difficulties are encountered in the preparation of gradients of larger volume (20–50 ml). Apparently, the change in hydrostatic pressure in the burette and the compressibility of the air in the space above the fluid level of the mixing chamber

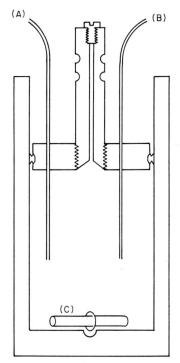

Fig. 24. Cross-section of mixing chamber recommended by Henderson. (From A. R. Henderson, *Analyt. Biochem.* **27**, 315 (1969), with permission.)

cause changes in the flow rates which make it difficult to keep the gradient volume constant within the required limits. Henderson recommends an improved mixing chamber in which the air space has been eliminated (Fig. 24).

a. Concave exponential gradients

Concave exponential gradients are formed if the initial concentration in the mixing chamber is larger than in the reservoir. Consequently, the concentration of the solution at the *bottom* of the gradient tube is given by the initial concentration (C_b) in the mixing vessel. Introducing the condition $C_i = C_b > C_R$ into Eq. (10) yields the expression describing concave exponential gradients:

$$C_v = C_R + (C_b - C_R)\, e^{-V/V_m}. \tag{10a}$$

It is evident that for a given gradient volume $V = V_g$, and for a given bottom concentration C_b, the top concentration $C_v = C_t$ is determined by the choice of the parameters C_R and V_m. Conversely, for any chosen pair of top and bottom concentrations, there is a family of gradients differing only by their curvature (Fig. 25). Thus, the *shape* of the curve is determined by the mixing volume: decreasing the mixing volume increases the concave curvature, whereas increasing the mixing volume

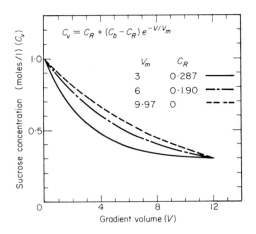

Fig. 25. Effect of changing parameters on shape of concave exponential gradients with common top and bottom concentration. With increasing mixing volume the gradients become flatter. In the example chosen, the mixing volume cannot be increased beyond 9·97 ml if a top concentration of 0·3M is desired, because the sucrose concentration in the reservoir cannot become < 0. Hence, of all concave exponential 12-ml gradients with a 0·3M top and 1·0M bottom concentration, the one represented by the top curve is the shallowest possible.

tends to make the gradient more nearly linear. It is also evident that if the same top concentration is to be reached with a larger mixing volume, the concentration of the solution in the reservoir (C_R) has to be reduced. In practice, the diluent added from the burette is often buffer without sucrose, so that $C_R = 0$ and Eq. (10a) takes the form

$$C_v = C_b\, e^{-V/V_m}. \tag{10b}$$

In this case V_m is fixed for a given top and bottom concentration and, hence, the shape of the gradient is no longer a matter of choice. As already mentioned in the discussion of linear gradients, the solution has to be introduced from the top of the gradient tube because the sucrose concentration is decreasing.

b. Convex exponential gradients

Convex exponential gradients are formed if the initial concentration in the mixing chamber is smaller than in the reservoir. Hence, the solution first introduced into the gradient tube will form the *top* concentration: $C_i = C_t < C_R$ and

$$C_v = C_R + (C_t - C_R)\, e^{-V/V_m}. \tag{10c}$$

Similar considerations apply as in the discussion of concave exponential

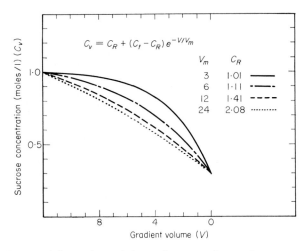

FIG. 26. Effect of increasing mixing volume on shape of convex exponential gradients.

gradients. For a given gradient volume V_g and top concentration C_t, the bottom concentration is determined by two variables, namely the sucrose concentration in the reservoir, C_R, and the volume in the mixing vessel, V_m. Hence, we can produce the same bottom concentration with different combinations of these two variables. It is easy to see that for a given set of top and bottom concentrations, the shape of the gradient is determined by the choice of these two variables. Decreasing the mixing volume increases the convex curvature of the gradients, whereas increasing the mixing volume tends to flatten the gradients (Fig. 26). At the same time, the reservoir concentration has to be decreased or increased so that the stipulated bottom concentration can be reached. Since the sucrose concentration increases during filling of the gradient tube, the outlet of the mixing vessel has to be introduced to the bottom of the gradient tube.

F. Analysis of Gradients

To exploit the analytical potential of zone velocity centrifugation, the methods applied to detect the components in the gradient should match the resolving power of the sedimentation process. This is rarely the case with discontinuous sampling methods which, in addition to the tedium, are subject to the cumulative errors from the many individual manipulations of counting drops and measuring the absorbancy of each

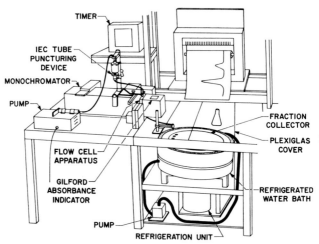

FIG. 27. Overall view of high-resolution gradient analyzing system. (From H. Noll, *Analyt. Biochem.* **27**, 130 (1969), with permission.)

fraction. Automatic scanning methods with suitable instrumentation are not only much more rapid and convenient but in addition are capable of reproducing the distribution of the macromolecules in the gradient without loss of resolution. In view of these advantages the costs for a high-resolution gradient analyzing system are moderate, since relatively inexpensive components can be combined with high-performance spectrophotometric equipment that is standard in most laboratories. Since these components are easily disconnected at the end of an experiment, the spectrophotometer remains available for its normal routine use.

An overall view of such an analyzing system which has been described in detail by Noll (1969) is depicted in Fig. 27. The centrifuge tube is seated in a puncturing device mounted on a Beckman DU monochromator which is part of a Gilford recording spectrophotometer system (Fig. 28). The tube is then punctured by depressing a spring-loaded lever that drives a needle up through the bottom of the centrifuge tube. The gradient drains at the negative hydrostatic pressure and constant rate maintained by the pump and passes through a flow cell

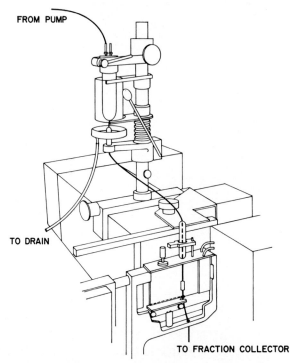

FIG. 28. Simplified drawing of puncturing device and flow cell holder with flow cell. Details of flow cell holder (see Fig. 30) have been omitted to give uncluttered view of essential features and of relative position of components with respect to each other. The Gilford absorbance indicator box in front of the cuvette compartment has been removed to reveal position of flow cell and holder and, in particular, the passage of the tubing out of flow cell through notch in sample changing carriage and hole drilled into bottom of compartment. (From H. Noll, *Analyt. Biochem.* **27**, 130 (1969), with permission.)

placed in the beam of the monochromator. Fractions of any desired size may be collected with the aid of an interval timer controlling the refrigerated fraction collector placed at the outlet end. Before puncturing the tube, a syringe filled with sucrose of a slightly higher density than the bottom of the gradient is connected at the outlet to flush out the air by filling the flow cell and flooding the needle. A commercial version of this gradient analyzing system is shown in Fig. 29.

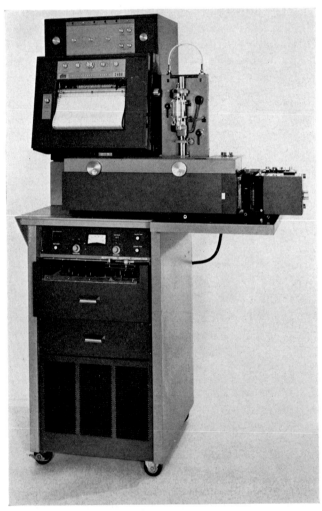

Fig. 29. Commercial version of automatic gradient analyzing system. (Courtesy International Equipment Company, Needham Heights, Massachusetts. (From H. Noll, *Analyt. Biochem.* **27**, 130 (1969), with permission.)

The flow cell is streamlined to reduce turbulence and has detachable quartz windows for easy cleaning. The cell is mounted in a holder designed to be placed into the cuvette compartment of the spectrophotometer (Fig. 30). Adjustment screws permit accurate alignment of the cell with respect to the center of the beam. An aperture with a slit that is adjustable from the outside confines the beam to a small area

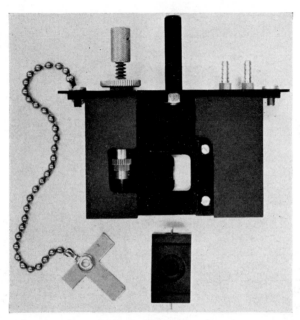

Fig. 30. IEC flow cell holder and, enlarged below, flow cell with removable quartz windows. (Courtesy, IEC.) (From H. Noll, *Analyt. Biochem.* **27**, 130 (1969), with permission.)

in the center of the flow stream. This is important in order to eliminate optical noise caused by schlieren resulting from slight turbulence at the boundaries of the stream.

A measure of the resolution attainable with this system is the resolution of 77S cytoplasmic and 73S mitochondrial ribosomes from *Neurospora crassa*, i.e. a resolution corresponding to a difference in sedimentation rate of only about 5% (Fig. 31).

G. Choice of Gradient

There is no general-purpose gradient ideally suited for all applications. The choice of rotor sizes and gradient shape should be dictated by the particular needs. If the principal objective is primarily analytical, the

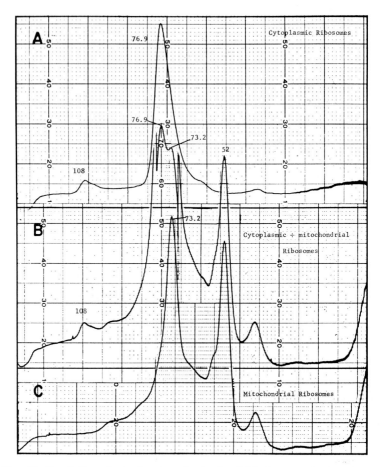

Fig. 31. Resolution of cytoplasmic and mitochondrial ribosomes from *Neurospora crassa* in an isokinetic gradient. Purified cytoplasmic (A) and mitochondrial ribosomes (C) have been mixed to show separation into two distinct peaks (B) Sedimentation was in an IEC rotor SB-283 for 3 h at 40,000 rev/min and 2°C. The isokinetic sucrose gradient, calculated for a particle density of 1·5, had a volume of 12·0 ml and a top concentration of 0·465M. The peak with an S value of 52 corresponds to the larger subunit of the mitochondrial 73S ribosomes. The breaks in the curve on either side of the 73/77S double peak indicate that the scale has been changed by a factor of two. (Unpublished results of H. Küntzel and H. Noll, taken from H. Noll, *Analyt. Biochem.* **27**, 130 (1969), with permission.)

very fast titanium rotors at present capable of spinning up to six samples at 60,000 rev/min are preferable because of the short centrifugation times and small sample quantities required. As the lengths of the buckets in the 60,000 rev/min rotors are shorter than in the slower rotors, the range over which high resolution is attainable is more limited. This becomes particularly serious if the mixture to be analyzed covers a wide spectrum of sizes, as in the case of ribosome preparations containing particles ranging from 30S subunits to polysomes of 500–600S. It is obvious that in this case the best overall resolution would not be obtained with an isokinetic gradient which displays the S-values on a linear scale, but rather with a steeper gradient in which the distance of travel corresponds to a logarithmic scale. This will allow the 30S and 50S subunits to move out of the top zone and separate from each other before most of the heavy polysomes have accumulated in the pellet. Conversely, if the aim is the separation of two objects with very similar sedimentation rates, sedimentation in isokinetic gradients will be the method of choice.

III. RELATIONSHIP BETWEEN AGGREGATE SIZE AND SEDIMENTATION CONSTANTS OF POLYSOMES

As we have already seen, an early clue to the structure of polysomes was the quantitative relationship between the number of ribosomes making up an individual polysome and its sedimentation coefficient. Since the sedimentation rates S_1, S_2 of particles of equal density in a centrifugal field increase as the surface/mass ratio is reduced, spherical particles of mass m_1, m_2 obey the relation

$$\left[\frac{S_2}{S_1}\right]^{\frac{3}{2}} = \frac{m_2}{m_1} \quad \text{or} \quad \frac{S_2}{S_1} = \left[\frac{m_2}{m_1}\right]^{\frac{2}{3}}. \tag{15}$$

This relationship is immediately derived from the volume/surface ratio of a sphere. If the aggregate is made up of n units ($m_2/m_1 = n$), and behaves hydrodynamically like a sphere, the sedimentation coefficient S_n of the n-fold aggregate would according to Eq. (15) assume the values

$$S_n = S_1 \times n^{\frac{2}{3}} \tag{16}$$

or

$$\log S_n = \log S_1 + \tfrac{2}{3} \log n. \tag{17}$$

In a log–log plot the values should fall on a straight line with a slope of 0·67 and an intercept of $\log S_1$. When this analysis is applied to polysomes, a straight line with a slope of 0·60 is obtained (Fig. 32). This shows that the frictional drag decreases somewhat less than expected for spherical aggregates. As a first approximation, Gierer (1963) suggested to treat the polysome as an ellipsoid corresponding to a linear chain of spherical 80S particles. Introducing into Eq. (16) the frictional coefficient $f(n)$ for an ellipsoid of axial ratio n, he obtained

$$S_n = 80\frac{n^{\frac{2}{3}}}{f(n)}. \qquad (18)$$

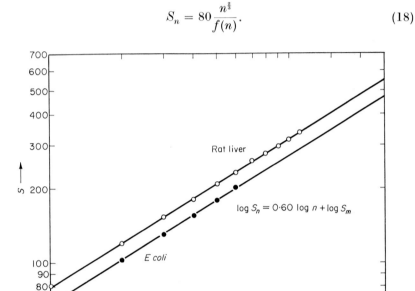

Fig. 32. Linear relationship between sedimentation coefficient and polysome size in log–log plot.

With this correction a satisfactory fit between measured and calculated values was obtained up to the trimer. For larger aggregates the experimental values were found to be higher, suggesting that the chains are not straight, but begin to form a more compact structure. Pfuderer, Cammarano, Holladay and Novelli (1965) showed that the data are compatible with a number of helical structures, a conclusion which has also received support from electron microscopic studies. It would be interesting to know to what extent ionic conditions and the average spacing between ribosomes influence the sedimentation rate of polysomes. As the ribosomes are spaced further apart, the frictional drag should increase approaching that of a random coil, unless there is a sufficiently strong messenger-independent interaction between ribosomes as in the case of the unspecific aggregates encountered at high Mg^{++} concentrations.

According to Fig. 32, the relationship between sedimentation coefficient and polysome size is very precisely determined by the equation

$$\log S_n = \log S_1 + 0\cdot60 \log n$$

and hence the number of ribosomes in a polysome sedimenting with a given rate is immediately given:

$$n = \text{antilog} \frac{\log S_n - \log S_1}{0\cdot60}.$$

This is particularly useful if we wish to know the size of the large polysomes present in the rapidly sedimenting unresolved regions of the gradient, e.g. the size of the structures corresponding to an S-value of 600. Thus, we immediately obtain

$$n = \text{antilog} \frac{2\cdot778 - 1\cdot903}{0\cdot60} = 29.$$

IV. RADIOASSAY METHODS

Almost all of our information on protein synthesis and on the structure and function of polysomes has been derived from a combination of the methods of ultracentrifugation already discussed and experiments involving the use of radioactive isotopes. In this section, our discussion will be limited to various techniques of studying polysome function by liquid scintillation counting since this has become the most versatile and efficient method of radioassay.

A. General Considerations

The basic problem encountered in one form or another in all work with water-soluble macromolecules is two-fold: (1) the separation of the labelled macromolecule from its low molecular weight precursors and (2) the dispersion of the radioactive product in an organic one-phase solvent system in which both the organophilic scintillators and the hydrophilic sample substance are soluble. In the simplest case, the separation of precursor and product is accomplished by precipitation with acid as both nucleic acids and polypeptides are generally insoluble at low pH. In other cases, the components of low and high molecular weight are separated prior to radioassay by gel filtration, electrophoresis or ultracentrifugation. Since in these cases precipitation is often not required, it is most convenient to assay directly the aqueous samples. For some applications involving zone-velocity centrifugation in sucrose gradients, direct radioactivity determinations of aqueous samples are not suitable because the small precursor molecules (amino acids, nucleotides) are insufficiently separated from some of the slower sedimenting macromolecular components. As the radioactivity of the precursor molecules frequently exceeds that of the macromolecular product by a large factor, the upper third of the gradient will be covered by the radioactivity of small molecules in the top zone. A good example is the release of completed polypeptide chains from polysomes during incorporation in a cell-free system. It is immediately clear from an inspection of the series of gradients in Fig. 33 that this process could not have been studied without separating the acid-insoluble radioactivity of the newly formed polypeptides from the soluble radioactivity of the ^{14}C-leucine precursor.

B. Direct Assay of Aqueous Samples

A number of solvent mixtures have been recommended for the assay of aqueous samples. Not all of these are equally suitable for the assay of samples from sucrose gradients because sucrose tends to separate as a syrup at the high concentration end of the gradient. Two solvent systems, one based on dioxane, another on toluene, have been found to give highly reproducible results with fractions from sucrose gradients. If dioxane is the solvent, a thixotropic gel and sufficient water have to be added to keep the sucrose in solution (Peckham and Knobil, 1962; Jost, Shoemaker and Noll, 1968). Toluene forms a gel with water when mixed with the detergent Triton X-100 (Patterson and Greene, 1965; Benson, 1966). The Triton-toluene gels are ideally suited for the assay of samples from sucrose gradients because they form a stable, completely transparent, homogeneous phase even at high sucrose concentration and show a considerably higher counting efficiency (less quenching) than the dioxane-cabosil* gels. Additional advantages are the greater ease of handling (the dioxane-cabosil thixotropic gels are rather messy) and the much lower cost. Table 3 gives the composition of the dioxane-cabosil and toluene-Triton X-100† mixtures that have given the best results in my laboratory in the direct assay of sucrose gradient fractions.

The chief attraction of the direct assay of sucrose gradient fractions is simplicity, speed, excellent reproducibility and economy. The speed with which the samples can be prepared for radioassay is an important consideration because it determines the total number of samples that can be processed in a working day and, hence, the number of gradients of a given resolution. Good resolution requires the collection of at

---

* Cabosil, No. M-5, Cabot Company.
† Triton X-100, Rohm & Haas.

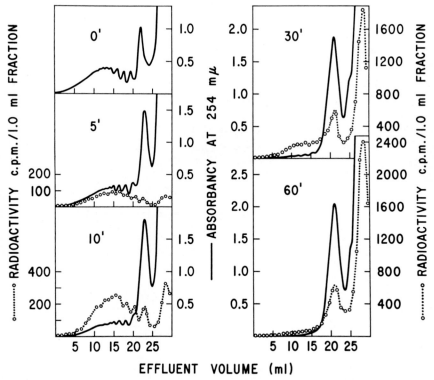

EFFLUENT VOLUME (ml)

FIG. 33. Sedimentation patterns illustrating various stages of polysome break-down, chain completion and release. Livers of 250 g rats were homogenized in 2·5 parts of medium S (0·0035M MgCl$_2$, 0·140M NH$_4$Cl, 0·03M Tris-HCl, pH 7·5) containing sucrose (0·11M) and 2-mercaptoethanol (0·005M). Crude post-mitochondrial supernatant (0·3 ml; equivalent to 0·085 g wet liver or 0·35 mg RNA, and containing 25 mμmoles of ^{12}C-leucine as determined by isotope dilution) was incubated at 25°C after adding the following components (μmoles/ml incorporation mixture): ATP (1·0), GTP (0·4), tricyclohexylammonium phosphoenolpyruvate (10), crystalline pyruvate kinase (10 μg), Tris-HCl, pH 7·5 (30), MgCl$_2$ (3·0), NH$_4$Cl (140), 2-mercaptoethanol (5), 19 ^{12}C-L-amino acids except leucine (0·06 each), and uniformly labelled ^{14}C-L-leucine (0·047, 10·7 mc/mole). At the end of the indicated periods, 1·0-ml samples were chilled, treated with 0·1 ml of 10% DOC to dissolve the microsomal lipoprotein membranes and layered in the form of an inverted gradient over 27·5 ml of a convex exponential sucrose gradient. The sucrose gradient contained Tris, MgCl$_2$ and NH$_4$Cl in the same concentrations as the incorporation mixture. Centrifugation was for 2·5 h (patterns on the left) or 3·0 h (patterns on the right) at 25,000 rev/min and 2–4°C. Counting efficiency was 68%, recovery of radioactivity from gradients 85–90%. (Unpublished results of H. Noll and T. Staehelin, taken from H. Noll, Polysome Organization as a Control Element, op. cit., with permission.)

least 30 fractions per gradient, high resolution as many as 60. Further increase of resolution will be limited by the combined errors of sampling and radioactivity measurements, a limitation inherent in all discontinuous methods of gradient analysis. In the case of the direct assay of sucrose gradients, both the speed and accuracy of sample preparation are greatly improved without much effort and at relatively low cost by automation. Thus, using the type of automatic gradient

TABLE 3

Composition of scintillation mixtures for direct assay of aqueous samples from sucrose gradients

		grams
A. Dioxane–Cabosil:		
Naphthalene (Fisher, recrystallized from alcohol)		100
PPO		7·0
POPOP		0·3
Cabosil		30
1,4-dioxane to make 1000 ml		
B. Toluene Triton 76:		
Mix toluene scintillator-A		700 ml
+Triton X-100		600 ml
Toluene Scintillator-A		
PPO—4 g		
POPOP—0·1 g		
toluene—1000 ml		

analyzing system discussed in Section IIF (Figs. 27–29), the effluent from the flow cell may be collected directly into the scintillation vials by means of the automatic fraction collector. The scintillation reagents may be added in the same operation by an automatic delivery system coupled to the indexing mechanism of the fraction collector. A Fisher* Volustat syringe pump designed for the repetitive dispensing of a fixed volume in the range from 1 to 12 ml has proved satisfactory for this purpose in the author's laboratory.

The collection of the very small fractions (0·5 ml and less), produced when the relatively small volumes (4–14 ml) of the high performance titanium rotors are cut into 30 equal fractions, presents serious difficulties with conventional fraction collectors in which the fluid is allowed to drop into a series of tubes at a constant rate. These difficulties arise whenever the drop size is no longer small in comparison

---

* Fisher Scientific Co., Pittsburgh, Pennsylvania, U.S.A.

to the fraction volume because (1) the desired fraction volume does not correspond to an integral number of drops, (2) drops are lost between tubes at the moment of changing fractions and (3) a relatively large and

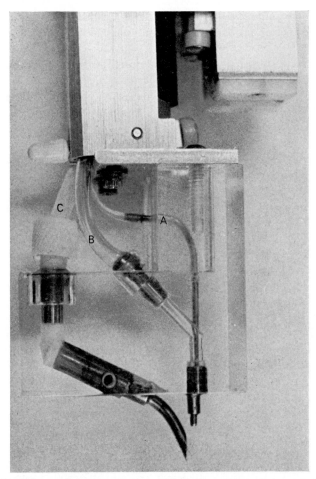

Fig. 34. Micro-collection adaptor with flushing tip and reagent dispenser. A. Sample stream from sucrose gradient; B. tubing leading to Volustat pump delivering distilled water for flushing sample delivery tip and C. Teflon tubing connected to Volustat pump adding scintillation reagents.

variable portion of the fraction volume remains at the tip of the delivery tubing. An adaptor suitable for attachment to any time-operated fraction collector was designed by Noll to eliminate these difficulties by flushing the delivery tip with a predetermined volume of water immediately before collection is advanced to the next tube. The adaptor (Fig. 34) consists of a Lucite block with a narrow vertical bore holding a No. 19 gauge needle that connects to the tip of the tubing from the gradient tube. The protruding tip of the needle is surrounded coaxially by a somewhat wider flushing tip at the end of a flushing channel which joins the delivery needle in a knee-like fashion from the side. The upper end of the flushing channel is connected to the flushing pump.

The adaptor shown in Fig. 34 was designed for the Savant Unifrac gradient fractionator.* In this ingenious device a delivery motor shuttles along a monorail track to deliver the effluent successively to each of 100 liquid scintillation vials arranged in a square tray. On command of an electronic interval timer in the control unit, the delivery motor very rapidly moves to the next vial. An event marker output on the control unit produces a 5 V pulse at the moment of indexing. This signal is used to drive an auxiliary timer which controls the flushing pump. The auxiliary timer is set for an interval slightly shorter than that of the master control unit so that the flushing occurs immediately before the delivery motor advances to the next vial.

An additional delivery spout on the adaptor is connected to a second syringe pump which is wired to the same auxiliary timer as the flushing pump and dispenses liquid scintillation fluid. A diagram of the circuit which adapts the signal from the event marker output of the Savant control unit to the input for the auxiliary timer and Volustat pumps is shown in Fig. 35. An overall view of the Savant Unifrac fraction collector modified for micro-sample collection and automatic reagent addition is reproduced in Fig. 36.† When combined with the Gilford spectrophotometer and IEC tube puncturing and flow cell components described in the preceding section (IIF), this assembly forms an integrated system capable of spectrophotometric scanning, collecting and processing sucrose gradients for scintillation counting in a completely automatic operation. A general view of this system is shown in Fig. 37.

By way of example let us consider a typical experiment. Reconstruction of the active complex in the $E.$ $coli$ protein synthesizing system requires $50S$ and $30S$ subribosomes, peptidyl-tRNA and mRNA (Jost et al., 1968). The assembly proceeds through the intermediate of an unstable mRNA-peptidyl-tRNA-$30S$ complex which is stabilized by the addition of $50S$ particles. The $30S$ complex may also be stabilized by the binding of streptomycin to the $30S$ particles (Jost et al., 1968).

---

* Savant Instruments, Inc., Hicksville, New York.

† Available from Molecular Instrumentation Company, 2665 Orrington Avenue, Evanston, Ill. 60201, U.S.A.

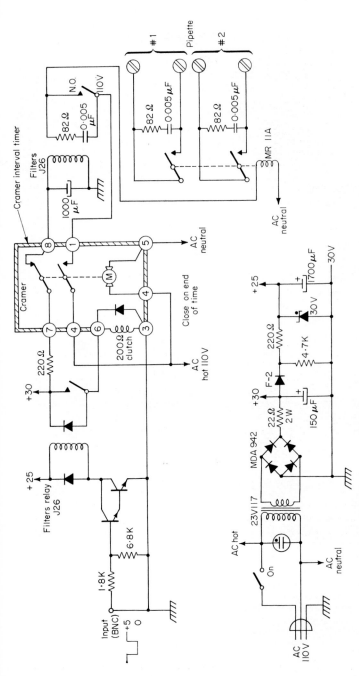

FIG. 35. Wiring diagram for slave timer. The 5-V signal of the Savant control unit (Fig. 36) is amplified by a power supply (schematics shown in lower portion of graph) to drive a *Cramer* interval timer which actuates the two Volustat pumps (upper portion of graph). The circuits were designed by Robert Loyd.

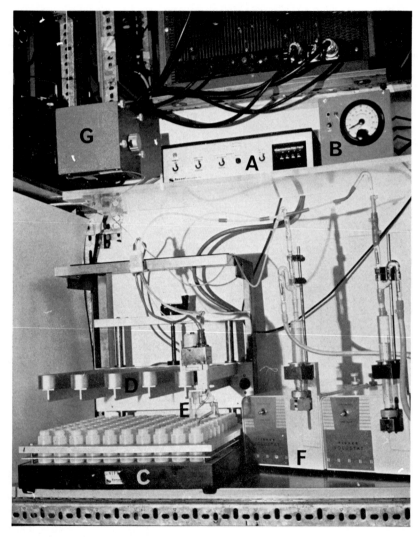

FIG. 36. Overall view of *Savant* Unifrac fraction collector modified for micro-sample collection and automatic reagent addition. A. master timer, B. slave timer, C. rack with scintillation vials, D. monorail tracks for engine of dispensing head, E. dispensing head with micro-collection adaptor, F. Volustat pumps with syringes, glass valves, and tubing leading to reservoir containing scintillation reagent, G. *Gilford* absorbency detector-indicator housing and flow cell compartment with tubing connecting flow cell to dispensing head of fraction collector.

To demonstrate the formation of the 30*S* and 70*S* complex, the reaction mixture consisting of poly A, ^{3}H-polylysyl-tRNA and purified subribosomes was analyzed by zone-velocity sedimentation in a sucrose gradient. Complex formation manifests itself by the appearance of radioactivity in the 30*S* and/or 70*S* peak positions as well as by the corresponding shifts in the absorbancy pattern. Thus, after incubation of

Fig. 37. Integrated system for automatic analysis of sucrose gradients with respect to absorbency and radioactivity. At left are the components of the spectrophotometric high-resolution gradient analyzing system illustrated in Fig. 27, at the right the automatic system (shown in detail in Fig. 36) which processes the gradient for radioassay by liquid scintillation counting. A. *Harvard* syringe pump, B. *Disk* automatic digital printer for peak integrator attached to Honeywell strip chart recorder, C. amplifier of *Gilford* spectrophotometer, D. *Beckman* monochromator, E. *Gilford* absorbency detector-indicator housing, F. *Honeywell* recorder, G. *IEC* tube-puncturing device.

50*S* ribosomes in the presence of excess poly A, 30*S* particles and polylysyl-tRNA, a new 70*S* absorbancy peak appears, evidently at the expense of the two slower sedimenting peaks, for the 30*S* peak has decreased and the 50*S* peak has vanished altogether. These changes are accompanied by the expected shift of radioactivity from the top of the gradient into the 70*S* peak position (Fig. 38).

In these experiments six gradient tubes containing 3·9 ml were centrifuged for 2 h at 60,000 rev/min and each gradient cut into 30 fractions after passage through the spectrophotometric scanning

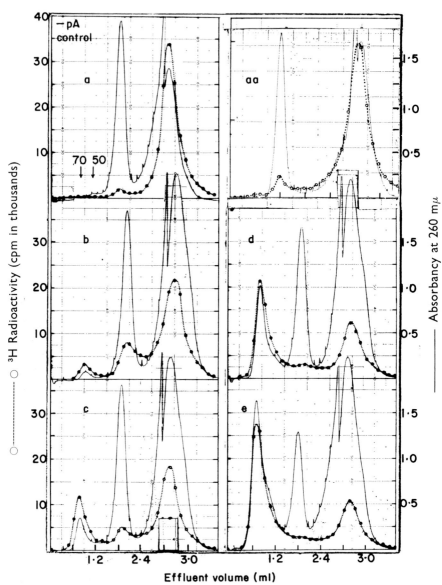

Fig. 38. Reconstruction of the active complex from *E. coli* subribosomes, poly A
and polylysyl-tRNA. Purified 30*S* particles were incubated for 20 min at 37°C
with an excess of ³H-polylysyl-tRNA and poly A and with increasing concentrations
of 50*S* particles (b–e). A control lacking poly A is shown in (a), another control in
which 50*S* particles have been used instead of 30*S* ribosomes is shown in (aa). Note
that in the complete system containing an excess of all other components, all the
50*S* particles combine with the equivalent amount of 30*S* subunits to form 70*S*
particles as evident from the shift in the 50*S* absorbency peak (compare for example
aa and d). The formation of 70*S* particles manifesting itself by the shift in absorbency
is accompanied by a corresponding shift of radioactive peptidyl-tRNA into the 70*S*
position. (From M. Jost, N. Shoemaker and H. Noll, *Nature, Lond.* **218** (1968),
with permission.)

system. Pumping at a rate of 0·6 ml per minute, the processing of all six gradients takes about 90 min. At the end of that time all of the 180 fractions will be ready for scintillation counting, if the automatic system described above is used.

C. Precipitation of Protein and Nucleic Acids with Acid

The standard method for measuring the incorporation of the monomeric precursors into polypeptides or polynucleotides is precipitation of the reaction product with acid, followed by washing procedures which ensure the complete removal of the soluble monomeric components usually present in large excess. This is usually followed by washes with organic solvents to remove radioactivity incorporated into phospholipids and lipoproteins. The separation of precipitate from supernatant fluid may be accomplished by centrifugation or filtration methods. Centrifugation methods are in general more time-consuming and precipitates are difficult to recover without loss. In the most widely used filtration methods, the solution containing the precipitate is passed with suction through a filter disk mounted on a fritted glass support base by means of a glass funnel clamped down onto the rim of the base. A commercially available filter holder of this type is the apparatus of Millipore*. If sucrose gradients are to be processed by acid precipitation, it is advisable to set a portion of high bench space in the laboratory aside for a battery of 30 suction flasks equipped with a Millipore type filter apparatus. The suction flasks may be connected in series of 10–15 each (depending on the available water pressure) to a 2 liter flask which serves as a trap and is connected to a suction pump (Fig. 39). A stopcock should be inserted between each suction flask and the T-tube connecting it to the common vacuum line. This is necessary in order to prevent the vacuum from breaking down after the filtration process has been completed in one flask and to regulate the vacuum applied to individual flasks which often have variable rates of filtration.

The main difficulty in collecting acid-insoluble precipitates by filtration is clogging of the filter disks. This reduces the efficiency of washing and thus explains the considerable scattering often encountered in the literature. The importance of complete removal of acid-soluble precursors, especially at low levels of incorporation, is appreciated if we consider that the radioactivity of the precursors is usually present in large excess. Thus, if the total input of precursor activity is 10^6 cpm, the efficiency of washing must be greater than 99·995% corresponding

---

* Millipore Corporation, Bedford, Massachusetts, U.S.A., Catalogue No. XX10-025-00.

to a blank value of less than 50 cpm in order to be able to measure incorporation activities of the order of 100 cpm. This means that precipitates must be very finely divided so as not to give rise to entrapment of acid-soluble precursors.

The material used for filtering is rather critical. To ensure quantitative retention of the extremely fine precipitates produced at low concentrations of the acid-insoluble macromolecules, membrane filters

FIG. 39. Filter battery used in preparing nucleic acid and protein samples for radioassay by precipitation with acid.

have been widely used. However, reproducible results with membrane filters are obtained only if the total amount of precipitable material does not exceed 200 μg (F. O. Wettstein, personal communication). Another argument against the use of membrane filters is the high cost. A method which employs a hardened, acid- and alkali-resistant cellulose filter paper (Whatman No. 50) and gives quantitative recoveries with negligible blank values over a wide input range has been developed by Wettstein et al. (1963). The principle of the method is to precipitate the macromolecules onto the surface of a finely dispersed carrier which by increasing the surface area of the precipitate enhances the efficiency of washing and rate of filtration. Purified diatomaceous earth (Johns Mansville's Celite) was found to be ideally suited for this purpose. In practice, the Celite is suspended in the perchloric acid solution used for precipitation. After adding the suspension to the solution for precipitation, the mixture is filtered with suction through a preformed layer of Celite. The Celite cushions, which are formed by suctioning equal portions of an aqueous Celite suspension through the filter disks, prevent the radioactive precipitates from coming into direct contact with the filter paper disks. After rinsing with organic solvents, the dried Celite pellets are transferred into scintillation vials with or without the filter paper disk (Table 4).

TABLE 4

Preparation of samples for scintillation counting by Celite-carrier method

1. If the volume of the incubation mixtures or of the fractions from sucrose gradients is smaller than 1·0 ml, bring it to 1·0 ml with H_2O.
2. Add 0·1 ml of 2% serum albumin solution to serve as co-precipitant (only necessary if sample contains less than 1 mg protein).
3. Precipitate with 1·0 ml 0·6N perchloric acid containing 20 mg/ml Johns Manville Celite. (Shake Celite suspension before addition.)
4. Overlay Whatman No. 50 filter paper disks with 10–20 mg Celite by filtering with suction the appropriate volume of an aqueous suspension.
5. Collect acid precipitates on disks with suction.
6. Wash precipitates four times with 5 ml of *ice-cold* 0·2N perchloric acid.
7. Wash with 8 ml propanol–ether (1 : 2).
8. Wash with 6 ml propanol–ether–chloroform (1 : 1 : 1).
9. Dry in air stream for 10 min and transfer pellet into scintillation vials.
10. Dissolve protein in 1·0 ml hyamine solution.
11. Add 15 ml toluene scintillator.

Note: The lipid solvents used for washing may be modified to suit the particular protein under investigation. The solvents recommended here were chosen because albumin (which was used as test material) was found to have appreciable solubility in the more commonly used ethanol–ether solvent. For the experiments with hemoglobin illustrated in Fig. 40, the precipitation of globin was carried out with acid acetone (1 ml conc. HCl per 30 ml acetone) instead of perchloric acid in order to split off the acetone-soluble hemin.

For optimal counting efficiencies the protein or nucleic acid precipitates must be dissolved in the scintillation liquid. Organic bases dissolved in non-aqueous solvents are best suited, because they render the precipitates soluble in toluene. Molar solutions of hyamine in methanol (p-(diisobutyl-cresoxyethoxyethyl) dimethylbenzyl-ammonium hydroxide) have been widely used for this purpose. Additional solubilizers (e.g. Soluene-100 (Packard) and NCS Solubilizer (Nuclear Chicago) that have been developed more recently and for which better properties, especially less quenching, have been claimed) are now coming into use.

The results of a linearity test of the Celite carrier method are shown
in Fig. 40. Samples covering the range from 1–20 μl were removed from
a reticulocyte lysate that had been incubated at 37°C for 10 min in
the presence of an ATP-generating system and a mixture of ^{14}C-amino
acids. In this case the precipitation was carried out with a suspension
of Celite in acetone–HCl in order to keep the heme in solution for if the
heme is precipitated with the globin, the quenching caused by the

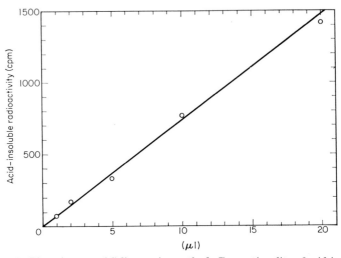

Fig. 40. Linearity test of Celite carrier method. Proportionality of acid-insoluble
radioactivity recovered to input of radioactive hemoglobin. The input was a 1 : 3
reticulocyte lysate containing about 70 μg/μl of hemoglobin. (Unpublished results
of Christoph Noll and Douglas Olsen.)

pigment gives rise to erratic values. The subsequent washing steps
followed the standard procedure detailed in Table 4. It is evident from
Fig. 40 that the acid-insoluble radiaoctivity recovered from the filters
is proportional to the input within less than $\pm 5\%$. The fact that the
straight line goes through the origin shows that there is no measurable
contamination by acid-soluble radioactivity.

Because of their finely dispersed state, precipitates collected with
Celite may be counted directly in toluene without solubilization by
means of an organic base. If the isotope to be counted has a radiation
energy corresponding to that of ^{14}C or greater, the reduction in counting
efficiency in comparison to samples dissolved in hyamine will be small
and more than compensated by the savings with respect to time and
reagent costs. However, precipitates labelled with ^{3}H require solubilization because otherwise most of the weak radiation energy of this
isotope would be absorbed within the precipitate.

The efficiency of precipitation depends on the nature and concentration of the macromolecules and on the precipitating agent. In order to ensure complete recovery by precipitation from solutions which contain very low concentrations of the polymer under study, an excess of carrier protein (usually bovine albumin) or nucleic acid is frequently added. Perchloric acid (PCA) and trichloroacetic acid (TCA) are the standard reagents for the precipitation of nucleic acids and proteins. Unfortunately no systematic studies seem to exist which would provide a rational guidance for the best choice in a given application. For the precipitation of nucleic acids, PCA is preferable because it hydrolyzes phosphodiester linkages less rapidly in the cold than TCA. The precipitability of polypeptides is a function of chain length and the ratio of hydrophilic to hydrophobic side chains. This is illustrated dramatically by the behavior of the homopolymers polyphenylalanine and polylysine. Oligopeptides of phenylalanine are extremely insoluble in acid, whereas even long chains of the basic polylysine cannot be precipitated with either PCA or TCA. However, Na-phosphotungstate, either alone or in combination with TCA (Gardner *et al.*, 1962), has been found to be effective for the precipitation of polylysine (cf. article by A. von der Decken in Vol. 1 of this series). It should also be recalled that TCA is extractable with ether or chloroform while PCA forms a water-insoluble potassium salt, both of which are useful properties whenever the removal of these precipitating agents is desired.

A filter disk precipitation method for the rapid processing of numerous samples of small volume has been described by Mans and Novelli (1961). In this method the aqueous samples are added by sorption to filter paper disks that have been numbered with pencil and impaled on steel pins stuck into a foam board. The macromolecules are then precipitated into the matrix of the fiber by immersing the disks in 10% TCA. Subsequently, the disks are processed batchwise through a succession of washes in acid and organic solvents (Table 5). The method was originally developed for following the time course of the incorporation of amino acids into protein in cell-free systems. For this reason, one of the steps calls for hydrolysis of charged tRNA by heating the disks at 90°C in TCA. At the end of the washing cycle, the radioactivity of the disks is counted by immersion in a toluene scintillator. The upper limit of sample volume that can be handled by this method is determined by the sorptive capacity of the disks. The capacity of heavy 3 MM Whatman disks of 25 mm diameter (which corresponds to the inside diameter of most scintillation vials and thus ensures counting with reproducible geometry as the disk rests flat on the bottom) is about 0·12 ml. The authors showed that, within the sorption capacity, the

radioactivity was proportional to the amount of protein added up to
0·8 mg protein per 23 mm disk, regardless of the concentration in which
the protein was applied. About 80% of the protein could be extracted
from the disks with 80% formic acid.

TABLE 5

Preparation of samples for radioassay by filter disk method

Pipette between 1 and 100 μl on 3 MM Whatman filter disk (2·1 cm diameter)
allow fluid to soak in for 5 sec.

$\downarrow$

1.	10 min	10% TCA	1°C

$\downarrow$

| 2. | 5 min | 5% TCA | 1°C |

$\downarrow$

| 3. | 30 min | 5% TCA | 90°C |

$\downarrow$

| 4. | 15 min | 5% TCA | 1°C |

$\downarrow$

| 5. | 10 min | ether–ethanol (1 : 1) | 25°C |

$\downarrow$

| 6. | 5 min | ether–ethanol (1 : 1) | 25°C |

$\downarrow$

| 7. | 10 min | ether | 1°C |

$\downarrow$

| 8. | 5 min | ether | 1°C |

$\downarrow$

air dry and put in 10 ml of toluene scintillator.

Note: Disk may be stored indefinitely at step 1. The background of radioactivity not
bound to protein and remaining after each washing step will be a direct reflection of the
volume dilution factor. Thus, the wash volume and number of washing steps required
depends on the level of background that is acceptable in a particular experiment. For
the experiments with hemoglobin illustrated in Fig. 41, the disks were first washed by
stirring for 10 min in acetone–conc. HCl (30 : 1) in order to remove hemin.

Although the authors recommend that the disks be dried for 15 sec in
a stream of warm air, tests in my laboratory revealed that the reproducibility is greatly improved if the samples are simply allowed to soak into
the paper for no more than 5 sec before dropping them into the acid
bath. Apparently, rapid drying in an air stream causes non-uniform
distribution of the protein with local crusting which entraps acid-
soluble radioactivity. When the filter disk technique was subjected to
a linearity test, the values fell on a straight line with very little scattering. However, in contrast to the Celite carrier method, with which it
was compared under identical conditions (Fig. 41), the line failed to go
through the origin, as already pointed out by Mans and Novelli (1961).

The contamination with acid-soluble radioactivity reflected in the intercept on the ordinate corresponds to the dilution factor in the batchwise washing procedure. It can also be seen that with ^{14}C the counting efficiency is slightly higher in the Celite carrier method. However, in most cases involving small volumes, the filter disk method will be preferred because it is less time-consuming and does not require expensive solubilizers for radioassay in a toluene scintillator.

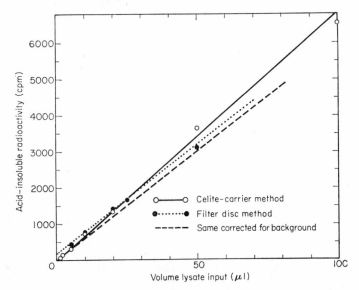

FIG. 41. Comparison of Celite carrier and filter disk methods for radioassay of radioactive hemoglobin. (Unpublished results of C. Noll and D. Olsen.)

The filter disk method may be modified to suit particular needs. In the linearity tests shown in Fig. 41, hemoglobin synthesized in a cell-free system was used. In order to eliminate the strongly quenching pigment, the filter disks were first washed with acetone–HCl. The heating in TCA was omitted because the contribution of charged tRNA to the total acid-insoluble radioactivity was negligible under the conditions of the experiment.

V. PREPARATION OF POLYSOMES FROM VARIOUS SOURCES

A. Mechanism of Polysome Breakdown

The main difficulty encountered in the isolation of polysomes is their fragility. There are four major mechanisms by which polysomes are known to break down: (1) cuts by endonuclease action in the messenger

7

thread linking the ribosomes together, (2) mechanical disruption of mRNA by shearing forces, (3) dissociation of ribosomes from mRNA at low Mg^{++} concentrations and (4) metabolic release of ribosomes under

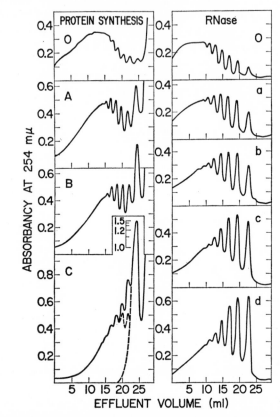

FIG. 42. Kinetics of polysome breakdown during protein synthesis (n − 1 process) and during incubation with pancreatic RNase (random fragmentation). Left: polysomes were incubated under conditions of protein synthesis at 25°C for 0 (O), 5 (A), 10 (B) and 30 (C) min and then analyzed by zone-velocity centrifugation. Right: polysomes (1·0 mg RNA/ml) were incubated at 30°C in the presence of 10^{-9} g/ml of crystalline pancreatic RNase for 1 (a), 3 (b) and 8 (c) min. Sample (d) was incubated for 8 min at an RNase concentration of 1·5 × 10^{-9} g/ml. The sedimentation pattern of the original untreated polysomes is shown at the top (O). (H. Noll, T. Staehelin and F. O. Wettstein, *Nature, Lond.* **198**, 632 (1963), with permission.)

conditions which inhibit chain initiation relative to chain completion. Since efforts to improve a given procedure depend on the correct diagnosis of the trouble, it is necessary to discuss the characteristic features of these different breakdown mechanisms.

Enzymatic and mechanical fragmentation proceed both by random scission of the messenger strand and, therefore, exhibit similar breakdown kinetics which are quite different from the kinetics of mechanisms (3) and (4) which both involve a separation of ribosomes or subribosomes from intact messenger strands. Thus, procedures resulting in the random scission of the messenger proceed through stages marked by the accumulation of a spectrum of polysome fragments of gradually diminishing average size (Fig. 42a–d), whereas treatments causing the separation of individual ribosomes ($n - 1$ process) manifest themselves by an immediate and conspicuous increase of the monomer fraction, accompanied by the gradual disappearance of the polysomes from the heavy end of the spectrum (Fig. 42A–C). Since both processes finally result in the complete conversion of polysomes into single ribosomes, they are difficult to distinguish in the very late stages.

A rigorous quantitative test distinguishing between the two decay mechanisms may be based on the feature that in the $n - 1$ breakdown both the total particle number (polysomes + monomers) and the number of monomers increase by one for every event (if we disregard the special case of the small fraction of dimers giving rise to two monomers), as illustrated below for the decay of a heptamer:

$$
\begin{array}{ccccccc}
 & 1 & 1 & 1 & 1 & 1 & 1 \\
 & \uparrow & \uparrow & \uparrow & \uparrow & \uparrow & \uparrow \\
\end{array}
$$

Size of breakdown $7\rightarrow6\rightarrow5\rightarrow4\rightarrow3\rightarrow2\rightarrow1$
products

No. of events   1   2   3   4   5   6

By contrast, in the polysome breakdown from random cuts in mRNA, not every cleavage produces a monomer:

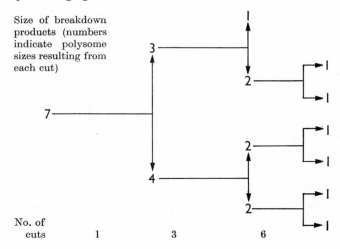

Size of breakdown products (numbers indicate polysome sizes resulting from each cut)

No. of cuts   1   3   6

and the larger the polysome, the less the likelihood that a random cut will produce a monomer.

Hence, in the first case, a plot of the increase of monomers against the increase in total particle population should give a straight line with unity slope; in the second case, a concave curve would be expected in which the increasing slope reflects the growing proportion of monomers produced during the process. The exact shape of the curve depends on the initial size distribution of the polysomes in the sample and may be computed by mathematically mimicking the process by a Monte Carlo method.

The examples of the two breakdown mechanisms shown in Fig. 42, i.e. the fragmentation of polysomes by low concentrations of pancreatic ribonuclease (in the range of $0 \cdot 001–0 \cdot 01$ μg/ml) and the release of ribosomes as a result of chain termination under conditions where chain initiation is blocked, have been analyzed quantitatively to test the two models. The results of a quantitative evaluation of the peak areas plotted in Fig. 43 show excellent agreement between the theoretical curves and experimental values and thus confirm more rigorously the validity of our concepts of the polysome structure.

The extreme sensitivity of polysomes to RNase is not surprising; since a single break of a phosphodiester linkage in the messenger strand may convert a hexamer for example into two trimers, this is obviously one of the most sensitive tests for RNase available. Single ribosomes prepared by methods involving messenger fragmentation remain blocked with a piece of stuck messenger which explains why they are not responsive to added messenger (Gierer, 1963; Staehelin et al., 1963). If RNase action is allowed to continue beyond the relatively few cuts required to convert polysomes into single ribosomes, the accessible portions of mRNA are further digested, leaving protected stretches of about 25–30 nucleotides buried within the ribosome structure (Takanami and Zubay, 1964; Takanami, Yan and Jukes, 1965).

Fragmentation of polysomes by mechanical shear is extensive when harsh methods are required to break cells. Grinding bacterial cells with alumina destroys polysomes almost completely. The French pressure cell, on the other hand, causes complete conversion to monomers and subunits at high pressure while yielding largely intact polysomes at low pressures (Zimmermann and Levinthal, 1967). Shearing forces resulting from repetitive pipetting also cause fragmentation of polysomes as shown in Fig. 4 (Oppenheim et al., 1968).

The $n - 1$ type of breakdown, characteristic of processes which lead to the separation of ribosomes from intact messenger strands, is observed under a variety of conditions. Most frequent are metabolic shifts, such

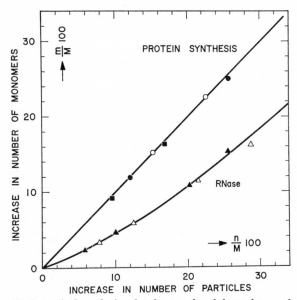

FIG. 43. Mathematical analysis of polysome breakdown by $n-1$ or random cleavage process. The two theoretical curves show the increase in number of monomers relative to the increase in number of particles (= monomers + polysomes) expected during the breakdown of a population of polysomes (a) by orderly stepwise monomer release from one end of each polysome (upper curve) and (b) by random cleavage of the messenger within polysomes (lower curve). The experimental points obtained under conditions (a) of protein synthesis and (b) incubation with RNase closely follow these theoretical expectations. The curve for random cleavage was computed by a Monte Carlo method as follows: a population of polysomes was represented by a population of sets of numbers in which successive cleavages were produced on the basis of a table of random numbers. The experimental values shown have been computed from several different experiments. (From H. Noll, T. Staehelin and F. O. Wettstein, *Nature, Lond.* **198**, 632 (1963), with permission.)

as from feeding to starvation, which cause a change in the relative rates of chain initiation and chain release (Noll, 1965; Staehelin, Verney and Sidransky, 1967). A very dramatic breakdown occurs upon cooling of intact cells, apparently because chain initiation requires a much higher energy of activation than chain extension (Das and Goldstein, 1968). Consequently, as the temperature is reduced, the attachment of new ribosomes to mRNA rapidly comes to a halt, while chain completion and release of ribosomes from polysomes continue at a reduced rate. This effect is strikingly illustrated in Fig. 44 by the breakdown of polysomes observed during washing of reticulocytes in the cold and the reversal upon subsequent incubation of the cells at 37°C (H. Noll and C. Noll, to be published). It follows that prolonged exposure to low temperatures must be avoided if the objective is the isolation of

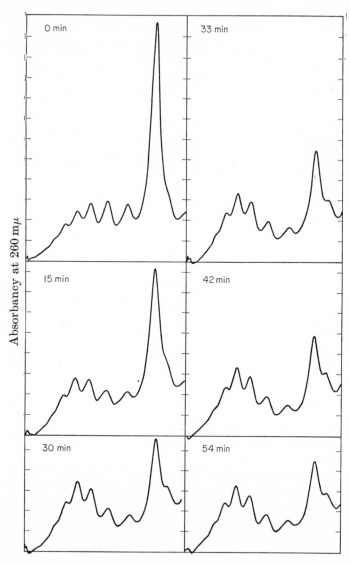

FIG. 44. Reassembly of reticulocyte polysomes at 37°C after breakdown during exposure of the cells to low temperature (~ 3°C). The cells were washed three times with cold buffer and then incubated at 37°C for 0, 15, 30, 42 and 54 min in a glucose salt buffer (140 mM NaCl, 1·5 mM MgCl$_2$, 5 mM KCl, 4 mM NaHCO$_3$ and 3 mM glucose) before lysis. (Unpublished results of Christoph Noll.)

polysomes in a state corresponding to their original size distribution. In some cases it might be advantageous to incubate the washed cells in the presence of glucose before homogenization in order to test whether a reversal towards larger polysomes takes place. Separation of ribosomes from mRNA is also encountered when polysomes are exposed to conditions which favor dissociation of ribosomes into subunits. This occurs if the Mg^{++} concentration is too low or the monovalent ion concentration too high, or as a result of fraudulent chain termination in the presence of puromycin (Villa-Trevino, Farber, Staehelin, Wettstein and Noll, 1964; Nathans, 1964; Noll, 1965; Schlessinger Mangiarotti and Apirion, 1967). An interesting pattern of breakdown combining features of both the messenger fragmentation and $n - 1$ type release process is encountered when polysomes are subjected to sonic oscillation. Apparently, the sonic vibrations have the dual effect of shaking ribosomes off the messenger strand as well as inducing breaks in the messenger (H. Noll, unpublished results). The patterns are difficult to reproduce, mainly because the commercial sonicators cannot be tuned reproducibly with respect to energy output.

B. Isolation of Polysomes

Unfortunately, there is no simple recipe applicable to the isolation of polysomes from any source. Each tissue or cell type presents its own peculiar difficulties and the optimal conditions have to be worked out by trial and error for each individual case in an often tedious process. The following general guidelines and technical pointers may be helpful:

(i) The methods of cell-breakage and homogenization must avoid application of high shearing forces.

(ii) The ionic environment is rather critical and the optimal conditions vary somewhat from case to case; as a rule the optimal concentration ratio of monovalent ions (K^+, NH_4^+) to Mg^{++} ranges from 10–20, and the Mg^{++} concentration from 0·003–0·01M. Bacterial-type ribosomes, including those of chloroplasts and mitochondria, require much higher Mg^{++} concentrations for stability (about 0·01M) than ribosomes from the cytoplasm of higher plants and animals (0·003–0·005M). At higher Mg^{++} concentrations ribosomes tend to form aggregates which, upon centrifugation through a sucrose gradient, produce patterns of a deceptively polysome-like appearance. However, the unspecific nature of these aggregates is revealed by their resistance to conversion into single ribosomes by traces (0·01 $\mu g/ml$) of crystalline bovine pancreatic RNase.

(iii) The greatest hazard during isolation is the extreme sensitivity of polysomes to RNase. A number of measures may be taken to reduce nuclease action, such as shortening the time of fractionation, working in the cold, avoiding the release of RNases by the rupturing of lysosomes and addition of various RNase inhibitors to the homogenate. Certain tissues are so rich in nucleases that isolation of polysomes has proved to be very difficult if not impossible. It is not yet certain whether addition of RNase inhibitors will provide a general method to protect polysomes from breakdown during isolation. However, it worked well with spleen homogenates from which only traces of degraded polysomes could be isolated unless RNase inhibitor from rat liver was added (Northup, Hammond and La Via, 1967). Bentonite is another RNase inhibitor that has been widely used, although mostly on faith and without convincing evidence of any beneficial effect. According to a recent report (Watts and Mathias, 1967), Bentonite is only effective when prepared according to exacting specifications which include selection of a certain particle size, removal of heavy atoms with EDTA and thorough elimination of residual EDTA. With Bentonite prepared according to these directions, the isolation of what appear to be undegraded polysomes has been possible from plant tissue which is notorious for its high content of nucleases.

(iv) The attachment to or interaction with membranes is another problem that often complicates the isolation of polysomes. Detergents have been valuable to overcome these difficulties, e.g. the use of deoxycholate to dissolve the endoplasmic reticulum or of Triton X100 to solubilize membranes of chloroplasts and mitochondria. In certain cases a combination of detergents and lytic enzymes has made it possible to open cells protected by tough envelopes without having to resort to mechanical rupturing methods which invariably destroy a large proportion of the polysomes. An example is the preparation of polysomes from bacteria that have been converted into spheroplasts with lysozyme and lysed with the non-ionic detergent Brij (Godson and Sinsheimer, 1967). This treatment also protects the polysomes from RNase, since the RNase is removed with the cell-wall during spheroplast formation.

(v) In some cases polysomes break down, not by mechanical damage, but metabolically by a preferential inhibition of chain initiation. This process may be prevented or retarded by adding, prior to homogenization, inhibitors of chain extension, such as chloramphenicol or actidione (cycloheximide).

1. Preparation of Polysomes from Rat Liver

In view of the foregoing discussions concerning the hazards involved in the preparation of polysomes, it is not surprising that almost a decade elapsed between the first preparations of ribosomes and the discovery of polysomes. In the case of rat liver, the isolation of polysomes was made possible by changes in the techniques of isolation which favored the selection of heavy aggregates and took advantage of the protective action of an RNase inhibitor present in the soluble fraction of the homogenate. Thus, Wettstein *et al.* (1963) observed that more active fractions were obtained when the ribosomes were released from the endoplasmic reticulum by adding the detergent deoxycholate directly to the postmitochondrial supernatant, instead of to the isolated microsomal fraction. The beneficial effect of this modification was later shown to be attributable to the presence of an RNase inhibitor in the post-mitochondrial supernatant (Bont, Rezelman and Bloemendal, 1965; Lawford, Langford and Schächter, 1966; Blobel and Potter, 1966). A similar RNase inhibitor was also demonstrated later in reticulocytes (Traub, Zillig, Millette and Schweiger, 1966) and, very recently, in *E. coli* (Smeaton and Elliott, 1967). The second modification introduced by Wettstein *et al.* (1963) was to spin the postmitochondrial supernatant through a 0·5/2·0M sucrose double layer. Under suitable conditions of centrifugation ($\omega^2 rt$), the pellet will contain mostly polysomes with almost no monomers because the viscosity barrier at the 0·5/2·0M interface retards the single ribosomes and a portion of the smaller oligomers sufficiently to prevent them from reaching the bottom. This method has proved generally useful for the isolation of polysomes and particularly in the case of homogenates that contain such a high proportion of single ribosomes that the presence of polysomes would be obscured. Sedimentation of the polysomes through 2M sucrose also results in a significant purification, since most of the proteins remain at the interface between the 0·5 and 2M sucrose layers.

The significance of these technical details becomes clearer from an inspection of the isolation procedure for rat liver polysomes illustrated in Fig. 45. Homogenization produces cell fragments covering a wide spectrum of sizes and morphological elements (Blobel and Potter, 1967). The large cell structures, i.e. nuclei, mitochondria and lysosomes, are then spun down at low speed, and the postmitochondrial supernatant, treated or not treated with DOC, is layered over a sucrose solution consisting of a 0·5M upper and 2M lower portion. During high-speed centrifugation in an angle-head rotor, the free polysomes will form a translucent pellet at the bottom of the centrifuge tube, while the membrane-bound polysomes, having a much lower overall density, accumulate at the 0·5/2·0M interface. Quantitative studies by Blobel and Potter have shown that the recoveries of bound ribosomes, which account for 75% of the cytoplasmic RNA, depend critically on the conditions of isolation. The yields are greatly improved by (a)

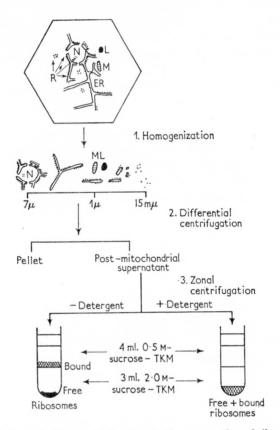

FIG. 45. Schematic diagram of the steps in the preparation of ribosomes from a postmitochondrial supernatant of rat liver. N, nucleus; M, mitochondrion; L, lysosome; ER, endoplasmic reticulum; R, ribosomes. (From G. Blobel and V. R. Potter, *J. molec. Biol.* **26**, 279 (1967), with permission.)

increasing the intensity of homogenization to prevent losses of large membrane fragments during low-speed centrifugation, (b) avoiding over-centrifugation during removal of nuclei and large cell particles (even under the minimal conditions necessary to remove the nuclei, about 35% of the total ribosomes are lost in this step) and (c) sufficiently vigorous centrifugation in step 3 to allow sedimentation of even the smallest ribosomal particles through the very viscous and dense ($d = 1 \cdot 27$ at 0°C) 2M sucrose layer.

2. Experimental Procedure

Male Wistar rats are starved for 16–18 h and then killed with a guillotine (Harvard Apparatus, Dover, Massachusetts). The livers (weighing about 10 g per 300 g animal) are removed as quickly as possible, blotted, trimmed of connective tissues and placed in tared 50 ml beakers containing 10 ml of ice-cold homogenizing buffer (0·25M sucrose; 20mM Tris, pH 7·5; 4 mM MgCl$_2$; 10 mM KCl; 1 mM 2-mercaptoethanol). After weighing, buffer is added to give a final

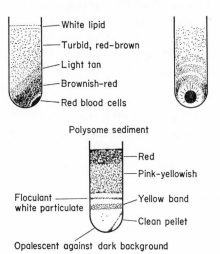

Sediment of postmitochondrial supernatant

— White lipid
— Turbid, red–brown
— Light tan
— Brownish–red
— Red blood cells

Polysome sediment

— Red
— Pink-yellowish
Floculant —— — Yellow band
white particulate
— Clean pellet

Opalescent against dark background

Fɪɢ. 46. Appearance of sediments in the preparation of postmitochondrial supernatant and purified polysomes from rat liver homogenates.

ratio of 2·5 ml/g liver. The livers are minced with scissors in the cold room and homogenized in a Potter–Elvehjem glass vessel with a motor-driven Teflon pestle (0·010 in. clearance), slowly at first to avoid heating, 6–8 up and down strokes after all large pieces have been disintegrated.

The homogenate is spun in 40 ml plastic Serval tubes at 13,000 rev/min (70% of top speed in Serval SS-1 centrifuge) for 10 min. Under these conditions cell debris, nuclei, mitochondria and lysosomes are sedimented into the pellet (Fig. 46). Approximately half of the opaque, red-brown supernatant above the multi-layered pellet is removed with a syringe from the middle of the tube. Care should be taken not to contaminate the post-mitochondrial supernatant (PMS) with material from the pellet or from the white lipid layer floating on the top. The needle on the syringe should be made of stainless steel. Other metals, especially those containing copper cause polysome breakdown by scission of the mRNA. Glass capillaries connected to the syringe through a rubber sleeve are equally suitable.

Addition of sodium deoxycholate (1.5 ml of a 10% solution per 10 ml PMS) to a final concentration of 1·3% in the postmitochondrial supernatant dissolves the membranes of the endoplasmic reticulum and produces a clear, dark red fluid.

The PMS may be examined directly for the polysome distribution pattern by centrifugation through a sucrose gradient as shown in Fig. 33, or it may be processed for the preparative isolation of purified polysomes (C-ribosomes) by centrifugation through a sucrose double layer as follows. Two-ml portions of ice-cold 2M sucrose in buffer (50 mM KCl, 20 mM Tris-HCl, pH 7·5, 5 mM MgCl$_2$) are dispensed into 10-ml polycarbonate screw-cap centrifuge tubes (IEC Catalogue No. 2067). An equal volume of buffered 0·5M sucrose is carefully layered over the bottom 2M sucrose layer. The tubes are then filled with the PMS and centrifuged for 1 h at 60,000 rev/min and 2°C (equivalent to 1·5 h at 50,000 rev/min or 2·5 h at 40,000 rev/min). Under these conditions the polysomes are selectively enriched in the pellet because most of the monomers and rapidly decreasing fractions of the small oligomers remain in the heavy sucrose solution. Complete recovery of the ribosomes requires four times longer centrifugation. The distribution of the components at the end of the centrifugation is illustrated in Fig. 46. The whitish particular matter that has precipitated at the interface of the sucrose double layer is probably DOC. The ferritin forms an intense yellow band below the interface. The pellets should be nearly colorless and completely transparent.

The supernatant is removed by aspiration with a capillary pipette and the sides of the tube are wiped clean with tissue paper. The pellets are then rinsed once again with several milliliters of cold buffer and dissolved by gentle agitation in 0·5 ml of distilled water or buffer. Dissolution is facilitated if the pellets are allowed to swell in water for about 30 min. If necessary, the solutions are clarified by low speed centrifugation. Good polysome preparations are opalescent; the light scattering disappears, however, if the polysomes are converted into single ribosomes with RNase. The purity of the polysomes is checked by measuring the absorbancy of a 1 : 200 dilution at 240, 260 and 280 mμ. The 260/280 ratio should approach 1·7. The sucrose gradient sedimentation pattern of a good preparation is shown in Fig. 11.

In a typical experiment 5 livers weighing 43·3 g gave 63·2 ml DOC-treated PMS which was distributed into 10 centrifuge tubes. The pooled resuspended pellets gave 4·8 ml of a solution with the following absorbancy readings in a 1 : 200 dilution: 0·245 (240 mμ), 0·410 (260 mμ) and 0·245 (280 mμ). Assuming that 10 Absorbancy units/ml = 10 mg ribosomes, this yield corresponds to 0·91 mg ribosomes per gram liver.

The polysomes may be stored in convenient portions (to prevent repeated freezing and thawing) at $-60°C$ without loss of activity.

3. Preparation of Polysomes from E. Coli

The isolation of relatively undegraded polysomes from bacteria required the development and perfection of methods for gentle lysis (Kiho and Rich, 1964; Dresden and Hoagland, 1965; Godson and Sinsheimer, 1967). This was accomplished by enzymatic degradation of the cell wall and subsequent lysis of the spheroplasts. As pointed out by Godson and Sinsheimer (1967), the reproducibility of the size distribution (manifested by the zone velocity sedimentation patterns) depends critically on the strict control of the experimental conditions. This includes conditions of growth and composition of growth medium as well as the detailed procedure of lysis. Rapid cooling of the cells is essential to preserve large polysomes because at lower temperatures chain initiation is more strongly inhibited than chain extension. This implies that formation of the initiation complex requires a much higher energy of activation than peptide bond formation and translocation (Das and Goldstein, 1968; Jost et al., (1968). As the cells grow older, they become increasingly resistant to conversion into spheroplasts by lysozyme. To obtain complete lysis, it is, therefore, necessary to harvest the cells before the density exceeds 2×10^8 cells per ml. The proportion of ribosomes present as polysomes depends on the growth rate (Godson and Sinsheimer, 1967): fast-growing cells in a rich medium contain relatively few subunits and $70S$ particles (Fig. 47A and B), while slower growing bacteria in synthetic or partially synthetic media contain a much larger fraction of $30S$ and $50S$ particles (Fig. 47C). It is interesting that the steady-state concentration of the $70S$ particles appears to remain fairly constant. These $70S$ particles appear not to be degradation products containing a fragment of mRNA because they increase greatly as a result of run-off, i.e. under conditions which inhibit initiation of new polypeptide chains more than the completion and release of nascent chains (Fig. 47D). The selective increase in $70S$ particles under conditions of chain termination argues against the idea suggested by Mangiarotti and Schlessinger (1967) that release requires dissociation into subunits. While it is well documented that free subunits can be converted into $70S$ particles by reducing the monovalent ion concentration relative to the Mg^{++} concentration, the converse seems not to be true, i.e. not all vacant (messengerless) $70S$ ribosomes will dissociate into subunits at a high $NH_4^+, K^+/Mg^{++}$ ratio. According to Subramanian, Ron and Davis (1968), a special factor is required to dissociate 70S particles that have been released from polysomes as a result of chain termination. Run-off may be greatly reduced by having chloramphenicol present in the medium during harvesting and sphero-plast formation (compare Fig. 47A and D). DNA has been reported to

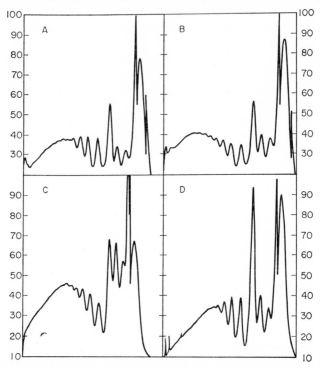

FIG. 47. Sedimentation patterns of *E. coli* polysomes prepared under different conditions. Polysomes from bacteria grown in tryptone broth: (A) lysates treated with high concentration of DNase (60 μg/ml lysate); (B) lysates treated with low concentration of DNase (0·6 μg ml lysate); (C) polysomes from bacteria grown in synthetic medium fortified with casein hydrolysate. All of these polysomes were prepared in the presence of chloramphenicol. (D) Polysomes prepared without chloramphenicol. (Unpublished results of P. Knolle.)

form complexes with polysomes (Byrne, Levin, Bladen and Nirenberg, 1964). If spheroplasts are lysed in the presence of DOC and DNase, 20–30% of the polysomes sediment with the DNA and membranes (Godson and Sinsheimer, 1967). Omission of DNase alone greatly reduces the yield of polysomes (P. Knolle, unpublished results). The concentration of DNase in the lytic mixtures has a marked influence on the resulting polysome pattern. As illustrated in Fig. 47, treatment with the relatively high concentration recommended by Godson and

Sinsheimer significantly reduces the proportion of heavy polysomes ($\leqslant$ heptamers) (Fig. 47A) in comparison to preparations produced at the hundredfold lower concentration of 0·6 μg per ml (Fig. 47B). It is not known whether this shift is the consequence of a contamination of the DNase preparation (although it was electrophoretically purified) with traces of RNase, whether it is the result of the degradation of membrane-bound DNA coincidentally sedimenting in this region, or whether it represents the fragmentation of DNA associated with nascent polysomes. As long as this question is not resolved, the lowest level of DNase compatible with good yields (0·5 μg/ml) is recommended for the standard procedure.

4. Experimental Procedure

The procedure follows essentially that of Godson and Sinsheimer (1967). The following stock solutions are prepared.

Tryptone broth: Separately sterilized solutions are added together: (a) 1000 ml containing 8 g DIFCO Bacto-Tryptone, 0·8 g yeast extract, 8 g NaCl + (b) 1 ml containing 0·1 g of glucose + 2 ml 1M CaCl$_2$. *TRIS*: Tris-HCl, 20 mM, pH 8·1. *Chloramphenicol* (CAM): 2 mg/ml in TRIS. *TMA*: 10 mM Tris-HCl (pH 7·5), 10 mM Mg-acetate, 100 mM NH$_4$-acetate. *Stock Solution A*: 45 ml TRIS, 50 ml sucrose (40% w/v in TRIS), 5 ml CAM. *Stock Solution B*: 5 ml sucrose (40% in TRIS), 0·4 ml CAM, 2·6 ml TRIS. Fresh for each experiment: *Lysozyme solution*: 1 ml containing 5 mg lysozyme in 2·5 mM EDTA and TRIS + 0·05 ml CAM. *Lysing solution*: 1 ml 100 mM MgSO$_4$ + 1 ml containing 50 mg Brij-58 (Atlas Chemical Industries) in 10 mM Tris-HCl, pH 7·5, + 2 ml containing 20 mg Na-DOC (Matheson) in 100 mM Tris-HCl, pH 8·1, + 0·25 ml 4M NH$_4$Cl + 0·25 ml CAM + 0·5 ml H$_2$O containing 5 μg DNase.

Cells of *E. coli* 3000 are grown overnight at 37°C with shaking in tryptone broth to which 1 mg thiamine has been added per 100 ml. The overnight culture is diluted 1 : 50 in thiamine-containing tryptone broth to give a final volume in the range of 10 to 200 ml and incubation is continued with shaking. The growth rates should be followed by counting the cells under the phase microscope. At a titer of 1×10^8 to 2×10^8 (after *ca.* 2–2·5 h of incubation) CAM is added (2·5 ml per 50 ml culture) and the cells are transferred into the cold room for harvesting and all subsequent manipulations are carried out as quickly as possible.

The culture flask is cooled to 2–4°C in an ethanol–ice bath (this should take less than 2 min). The cultures are transferred into 35-ml plastic tubes and centrifuged in a Serval SS-1 centrifuge at 6900 rev/min for 3 min. The pellets are suspended in 5 ml of A, transferred to 15 ml

COREX glass tubes and centrifuged as before. The pellet containing all the cells from the original culture is now suspended in 0·4 ml of B with the aid of a Vortex vibrator. Conversion into spheroplasts is initiated by adding 0·1 ml of lysozyme solution. The reaction should be complete in less than 60 sec; it manifests itself by a slight increase in viscosity. The lysing solution is then added (0·5 ml), and the reaction mixture carefully transferred to 2-ml centrifuge tubes. After 60 sec the lysate, which should have lost most of its turbidity, is centrifuged for about 3 min at 10,000 rev/min. The clear supernatant is decanted and either used immediately for sedimentation analysis or stored at −60°C. An aliquot (0·01 ml) is removed and diluted 1 : 100 with TMA to determine the absorbancy at 260 mμ. A reading of about 0·5 should be obtained from an input corresponding to 50 ml medium containing about 2×10^8 cells per ml.

REFERENCES

Benson, R. H. (1966). *Analyt. Chem.* **38**, 1353.
Berman, S. (1966). *In*, "The Development of Zonal Centrifuges and Ancillary Systems for Tissue Fractionation and Analysis", *Natn. Cancer Inst. Monogr.* **21**, 41 (U.S. Government Printing Office, Washington, D.C.).
Blobel, G. and Potter, V. R. (1966). *Proc. natn. Acad. Sci. U.S.A.* **55**, 1283.
Blobel, G. and Potter, V. R. (1967). *J. molec. Biol.* **26**, 279.
Bont, W. S., Rezelman, G. and Bloemendal, H. (1965). *Biochem. J.* **95**, 15c.
Britten, R. J. and Roberts, R. B. (1960). *Science, N.Y.* **131**, 32.
Byrne, R., Levin, J. G., Bladen, H. A. and Nirenberg, M. W. (1964). *Proc. natn. Acad. Sci. U.S.A.* **52**, 140.
Capecchi, M. R. (1966). *J. molec. Biol.* **21**, 173.
Das, H. K. and Goldstein, A. (1968). *J. molec. Biol.* **31**, 209.
Dintzis, H. M. (1961). *Proc. natn. Acad. Sci. U.S.A.* **47**, 247.
Dresden, M. and Hoagland, M. B. (1965). *Science, N.Y.* **149**, 647.
Engelhardt, D. L., Webster, R. E. and Zinder, N. D. (1967). *J. molec. Biol.* **29**, 45.
Gardner, R. S., Wahba, A. J., Basilio, C., Miller, R. S., Lengyel, P. and Speyer, J. F. (1962). *Proc. natn. Acad. Sci. U.S.A.* **48**, 2087.
Gierer, A. (1963). *J. molec. Biol.* **6**, 148.
Gilbert, W. (1963). *Cold Spring Harb. Symp. quant. Biol.* **28**, 287.
Godson, G. N. and Sinsheimer, R. L. (1967). *Biochim. biophys. Acta* **149**, 489.
Henderson, A. R. (1969). *Analyt. Biochem.* **27**, 315.
Jost, M., Shoemaker, N. and Noll, H. (1968). *Nature, Lond.* **218**, 1217.
Kiho, Y. and Rich, A. (1964). *Proc. natn. Acad. Sci. U.S.A.* **51**, 111.
Lawford, G., Langford, P. and Schachter, H. (1966). *J. biol. Chem.* **241**, 1835.
Mangiarotti, G. and Schlessinger, D. (1967). *J. molec. Biol.* **29**, 395.
Mans, R. J. and Novelli, G. D. (1961). *Archs Biochem. Biophys.* **94**, 48.
McCarty, K. S., Stafford, D. and Brown, O. (1968). *Analyt. Biochem.* **24**, 314.
McConkey, E. H. (1967). *Methods in Enzymology* **12A**, 620.
Nathans, D. (1964). *Fedn Proc. Fedn Am. Socs exp. Biol.* **23**, 984.

Noll, H. (1965). *In*, "Developmental and Metabolic Control Mechanisms and Neoplasia", 19th Annual Symposium on Fundamental Cancer Research published for the University of Texas M. D. Anderson Hospital and Tumor Institute, Houston, Texas, p. 67, The Williams & Wilkins Co., Baltimore.

Noll, H. (1967). *Nature, Lond.* **215**, 360.

Noll, H. (1969a). *Analyt. Biochem.* **27**, 130.

Noll, H. (1969b). *In*, "Protein Biosynthesis" (C. B. Anfinsen, ed.), Academic Press, New York and London (in press).

Noll, H. and Staehelin, T. (1969) to be published.

Noll, H., Staehelin, T. and Wettstein, F. O. (1963). *Nature, Lond.* **198**, 632.

Northup, R. V., Hammond, W. S. and La Via, M. F. (1967). *Proc. natn. Acad. Sci. U.S.A.* **57**, 273.

Oppenheim, J., Scheinbucks, J., Biava, C. and Marcus, L. (1968). *Biochim. Biophys. Acta* **161**, 386.

Patterson, M. S. and Greene, R. C. (1965). *Analyt. Chem.* **37**, 854.

Peckham, W. D. and Knobil, E. (1962). *Biochim. biophys. Acta* **63**, 207.

Penman, S., Scherrer, K., Becker, Y. and Darnell, J. E. (1963). *Proc. natn. Acad. Sci. U.S.A.*, **49**, 654.

Pfuderer, P., Cammarano, P., Holladay, D. R. and Novelli, G. D. (1965). *Biochim. biophys. Acta* **109**, 595.

Schlessinger, D., Mangiarotti, G. and Apirion, D. (1967). *Proc. Natn. Acad. Sci. U.S.A.* **58**, 1782.

Smeaton, J. R. and Elliott, W. H. (1967). *Biochim. biophys. Acta* **145**, 547.

Staehelin, T., Brinton, C. C., Wettstein, F. O. and Noll, H. (1963a). *Nature, Lond.* **199**, 865.

Staehelin, T., Wettstein, F. O. and Noll, H. (1963b). *Science, N.Y.* **140**, 180.

Staehelin, T., Wettstein, F. O., Oura, H. and Noll, H. (1964). *Nature, Lond.* **201**, 264.

Staehelin, T., Verney, E. and Sidransky, H. (1967). *Biochim. Biophys. Acta* **145**, 105.

Stent, G. (1963). Molecular Biology of Bacterial Viruses, p. 410, W. H. Freeman and Co., San Francisco, California.

Subramanian, A. R., Ron, E. Z. and Davis, B. D. (1968). *Proc. natn. Acad. Sci. U.S.A.* **61**, 761.

Takanami, M. and Zubay, G. (1964). *Proc. natn. Acad. Sci. U.S.A.* **51**, 834.

Takanami, M., Yan, Y. and Jukes, T. H. (1965). *J. molec. Biol.* **12**, 761.

Traub, P., Zillig, W., Millette, R. L. and Schweiger, M. (1966). *Hoppe Seyler's Z. physiol. Chem.* **343**, 261.

Villa-Trevino, S., Farber, E., Staehelin, T., Wettstein, F. O. and Noll, H. (1964). *J. biol. Chem.* **239**, 3826.

Warner, J. R., Knopf, P. M. and Rich, A. (1963). *Proc. natn. Acad. Sci. U.S.A.* **79**, 122.

Watts, R. L. and Mathias, A. P. (1967). *Biochim. biophys. Acta* **145**, 828.

Wettstein, F. O., Staehelin, T. and Noll, H. (1963). *Nature, Lond.* **197**, 430.

Zimmermann, R. A. and Levinthal, C. (1967). *J. molec. Biol.* **30**, 349.

CHAPTER 4

Fractionation of Ribosomal Proteins

HANS BLOEMENDAL AND CLAUDIA VENNEGOOR

Department of Biochemistry, University of Nijmegen
The Netherlands

I. INTRODUCTION

The study of protein biosynthesis is one of the most exciting topics in biochemistry at the present time. Although the different metabolic steps in this process are now rather well understood, the problem of the exact structure of the work-bench for protein production, the ribosome, is still far from being solved. The ribosome consists entirely of ribonucleic acid and a complex mixture of proteins. For the characterization of the latter, electrophoresis on starch gel and especially on polyacrylamide gel appears to be the most powerful tool at present, although chromatography on ion-exchange media has also been used. The present article attempts to survey procedures for the extraction and especially the separation of the protein moieties of ribosomes from different mammalian and bacterial sources and also from plants. In addition we include new data from our laboratory which have not previously been published.

II. TECHNIQUES

A. Preparation of Ribosomes

The isolation procedures for ribosomes and polysomes depend on the source from which they are derived. As a rough subdivision, however, two main methods can be distinguished. Bacterial cells may be ground

TABLE 1

Methods for the preparation of ribosomes used for the study of ribosomal proteins

Source of ribosomes	Reference
Escherichia coli	Tissières *et al.* (1959), Nirenberg and Matthaei (1961)
Rat liver	Rendi and Hultin (1960), Korner (1961), Cohn and Simson (1963), Wettstein *et al.* (1963)*, Bloemendal *et al.* (1964†, 1967†)
Rat kidney	Korner (1961)
Rat cardiac muscle	Rampersad and Wool (1965)
Rat skeletal muscle	Florini and Breuer (1966)
Calf lens epithelium	Bloemendal *et al.* (1966), Schoenmakers *et al.* (1967)
Rabbit reticulocytes	Schweet *et al.* (1958), Arnstein *et al.* (1964)
Dog pancreas	Keller *et al.* (1968)
MOPC-104 tumour in mice	Kedes *et al.* (1966)
HeLa cells	Warner (1966)
Nucleoli of HeLa cells	Warner and Soeiro (1967)
Plants (cytoplasm and chloroplasts)	Lyttleton (1968)
Pea seedlings	MacQuillan and Bayley (1966)
Neurospora crassa	Alberghina and Suskind (1967)
Yeast	Schmidt and Reid (1968)
Paramecium aurelia	Reisner *et al.* (1968)

† These references describe the preparation of polysomes from animal tissues without the use of detergents. This is in contrast to other methods for the isolation of polysomes quoted here, e.g. reference *.

with alumina when, as a rule, monosomes (70s ribosomes) are obtained. Plant and animal tissues, in contrast, are generally carefully disrupted in a glass homogenizer with a loose-fitting Teflon pestle. In this way it is possible to isolate polysomes which consist of 70s or 80s particles held together by a thin strand of messenger RNA. Table 1 provides references to a number of methods which have been used for the preparation of ribosomes from different species. Methods for the preparation of ribosomes have also been discussed by von der Decken (1967) and Bretscher and Jones (1967).

B. Preparation of Ribosomal Proteins

In order to isolate the proteins from ribosomes several procedures have been applied (Table 2). The yield of protein isolated per mg of ribosomal RNA depends not only on the extraction method but also on the source of the ribosomes.

TABLE 2

Media used for the extraction of ribosomal proteins

Extraction medium	Reference
67% Acetic acid	Waller and Harris (1961), Waller (1964), Traut (1966), Traut et al. (1967), Gesteland and Staehelin (1967), Möller and Widdowson (1967),* Terao et al. (1967), Moore et al. (1968)
4M LiCl and 8M urea	Low and Wool (1967)
3M LiCl and 4M urea	Spitnik-Elson (1965), Traut (1966), Krembel and Apirion (1968)
2M LiCl and 8M urea	Kedes et al. (1966)
2M LiCl and 5M urea	Lyttleton (1968)
2M LiCl and 4M urea	Leboy et al. (1965), Apirion (1966), Furano (1966), Alberghina and Suskind (1967), Gesteland and Staehelin (1967), Otaka et al. (1968), Reisner et al. (1968)
0·2N HCl	Cohn and Simson (1963), Cohn (1967)
Guanidinium chloride	Cox and Arnstein (1962), Cohn (1967)
Ribonuclease (endogenous)	Waller (1964), Spitnik-Elson (1962)
Ribonuclease (added)	MacQuillan and Bayley (1966), Gesteland and Staehelin (1967), Moore et al. (1968), Otaka et al. (1968)

* Extraction carried out in the presence of added ribonuclease.

It is important to note that ribosomes as normally isolated will have nascent protein still attached to them. Terao et al. (1967) have described a method for removing nascent protein from rat liver ribosomes prepared by the method of Rendi and Hultin (1960). The ribosomes are first treated with 15 mM EDTA and the 50s and 30s ribosomal sub-units separated by centrifuging on a linear sucrose density gradient (13·5–28·5% sucrose) at 25,000 rev/min for 8 h in the Spinco SW-25 rotor. After dialysis, the separate 50s and 30s sub-units are sedimented by centrifuging at 105,000*g* for 10 h.

The extraction methods summarized in Table 2 fall into three main categories:

1. The extraction may be performed with acetic acid. In this method two volumes of glacial acetic acid are added dropwise to one volume of the ribosomal suspension. The addition is performed over a period of 15 min with stirring. The mixture is allowed to stand for 2–4 h at 0–4°C after which the precipitated RNA is sedimented at 10,000*g* for 15 min. The supernatant is neutralized to pH 7·2 and freed from small ions by dialysis. Dialysis has to be stopped when the supernatant becomes cloudy. Ribonuclease may then be added to a concentration of 2 μg/ml. After incubation for 2 h at 0–4°C, dialysis must be started once more for a period of 12 h. The recovery of ribosomal protein by this method is about 80%.

2. The extraction is carried out by a combined treatment with urea and LiCl. Urea serves to dissociate the ribonucleoprotein, whereas LiCl serves to precipitate the RNA. To one volume of ribosomal suspension is added an equal volume of a solution of 8M urea containing 2–4M LiCl. After stirring for about 1 h, the mixture is kept overnight in the cold. The precipitate containing RNA and some protein is discarded after centrifuging and the supernatant dialysed as described in (1) above. This method also yields a recovery of about 80%.

3. In an alternative but less frequently used extraction method 0·25 ml of 2N HCl is added to the ribosomal pellet suspended in 2·25 ml of medium. The basic proteins are isolated from the supernatant obtained by centrifuging at 25,000*g* for 20 min. HCl and other small molecules are removed by dialysis against water. The proteins can finally be obtained by freeze-drying, precipitation with acetone at pH 8·0 or by precipitation with trichloroacetic acid.

According to Low and Wool (1967) the urea–LiCl procedure is more reproducible than the other extraction methods. Moreover, it provides better separation of protein from RNA. With pea seedlings the highest yield of protein was obtained when the extraction included treatment with ribonuclease in the presence of 8M urea (MacQuillan and Bayley, 1966). Radioactive ribosomal protein can be obtained easily from bacteria grown in a medium to which labelled algal protein hydrolysate is added, or from animals after intraperitoneal injection with a radioactive protein hydrolysate or pre-selected mixtures of labelled amino acids.

C. Electrophoresis on Starch Gel Containing Urea

Electrophoresis of ribosomal proteins on starch gels has been carried out mainly as described by Smithies (1955). However, several authors have applied modifications including 6M urea containing 0·0365M sodium acetate, pH 5·6 (Waller, 1964); 6M urea containing 0·02M sodium acetate, pH 4·2 (Terao et al., 1967); 1–6M urea containing 0·01–0·02N HCl, pH 2·05–3·1 (Cohn, 1967).

D. Electrophoresis on Polyacrylamide Gels

For the separation of proteins extracted from ribosomes in poly-acrylamide gels, several different systems are available. Thus a dis-continuous buffer system used in conjunction with gels which were polymerized with the aid of tetramethylethylene diamine (TEMED) at pH 4·5 (Reisfeld et al., 1962) has been adapted for electrophoresis of ribosomal proteins by the addition of 8M urea (Leboy et al., 1965). A continuous buffer system with gels containing 8M urea, 0·3M formic acid and 0·06M potassium formate at pH 3·5 has been used by Möller and Chrambach (1967). Warner (1966) and Warner and Soeiro (1967), in contrast, carried out the electrophoretic runs in a system described by Summers et al. (1965) with slight modifications. In this case the electrode buffers and the polyacrylamide gel contained 0·5% sodium dodecyl sulphate (SDS), 0·5M urea and 0·1M sodium phosphate, pH 7·2. In our laboratory the following continuous system with the same concentrations of urea and sodium acetate, pH 5·6, as used by Waller (1964) was used as a routine method. This method is now described in detail.

The following solutions were used:
1. 2·5 ml of a solution of acrylamide (30% w/v) and NN'-methylene-bisacrylamide (1·0% w/v).
2. 0·04 ml of β-dimethylaminopropionitrile.
3. 0·83 ml of 0·44M sodium acetate buffer whose pH had been adjusted to 4·6 at room temperature with acetic acid.
4. 3·44 ml of distilled water.

After mixing, the solution was carefully de-gassed in vacuo at the water pump and 0·5 ml of a freshly prepared solution of ammonium per-sulphate (2·4% w/v) was added. The components were mixed well and the solution poured into gel tubes (6·2 × 0·5 cm) and the surface layered with distilled water. Thirty min after mixing and pouring the solutions, the water layer was replaced with the upper electrode buffer solution consisting of 6M urea and 0·0365M sodium acetate whose pH has been adjusted to 5·6 with acetic acid. Complete polymerization of the gels was allowed to occur at room temperature for 18 h. The buffer in the lower electrode compartment contained 0·0365M sodium acetate, pH 5·6. The samples of protein were dissolved in buffer from the upper reservoir to give a final concentration of 1% w/v. Sucrose was then added to a final concentration of 15% w/v. Electrophoresis was carried out after application of the protein solution to the surface of the gel for 150 min at 80 V (6 mA per tube).

E. Detection of Ribonuclease Activity on Gels

Ribonuclease activity in the ribosomal protein can be demonstrated after the electrophoretic run in starch gel containing urea (Master and Srinivasa Rao, 1961; Waller, 1964). In our experiments ribonuclease activity was demonstrated on polyacrylamide gels by a modification of this method which is now described.

At the end of the electrophoretic run (in 5M urea at pH 5·6) the wet gels were cut longitudinally using a thin thread attached in a metal frame (Fig. 1). One half of the gel was then stained for protein using a solution of amido-black B. The other half was incubated for 1 h at 37° in 2·25 ml of a solution of 1% (w/v) RNA (sodium salt from *Torula* yeast, type II-S, Sigma) either in water or buffer. This RNA was previously purified by repeated precipitation from aqueous solutions with ethanol in the cold. Precipitation was facilitated by adding a few drops of concentrated NaCl solution. The precipitate was dissolved in water, freed from low molecular weight constituents by gel filtration on Sephadex G25 and subsequently freeze-dried. After incubating the gel in the RNA solution, the RNA which penetrated into the gel was precipitated by submerging the gel in 0·1N HCl. When the gels were incubated in 0·1% (w/v) solution of RNA, the penetrated RNA could be fixed permanently by staining with a solution of acridine orange (2% w/v) in acetic acid (15% v/v) containing lanthanum acetate (1% w/v), as described by Richards *et al.* (1965).

F. Fractionation of Ribosomal Proteins on Cellulose Columns

The acid-extracted proteins from ribosomal pellets can be purified on columns loaded with carboxymethyl cellulose. This material is prepared for use as described by Sober and Peterson (1962). According to Moore *et al.* (1968), 5–10 mg of protein per ml of packed column volume can be applied to a column having a length to cross-section ratio of 4 : 1. The protein is released from the column by gradient elution starting with 6M urea in 0·01M sodium acetate, pH 5·6, in the mixing chamber. There is introduced from a reservoir into the mixing chamber 6M urea in 0·5M sodium acetate, pH 5·6, until the effluent buffer is 0·2M in ionic strength. The reservoir in turn is supplied with sodium acetate solution to a final concentration of 1·35M. Elution is stopped when the effluent is 0·35M in acetate concentration. The best

Fig. 1. Device for cutting polyacrylamide gels longitudinally.

way to recover the proteins from the collected fractions is by dialysis against M acetic acid followed by freeze-drying. An alternative method involves precipitation of protein with trichloroacetic acid but some protein is lost due to its solubility in the trichloroacetic acid.

Fractionation of spleen ribosomal protein has been achieved on phosphorylated cellulose (Curry and Hersh, 1966). Phosphocellulose also has been used for the fractionation of *E. coli* ribosomal proteins (Traub *et al.*, 1967). In this method about 60 mg of protein are applied to a column (1·8 × 45 cm) which has been equilibrated with a buffer containing 6M urea and 3×10^{-3}M mercaptoethanol in 0·02M phosphoric acid neutralized to pH 7·0–7·3 with methylamine. Elution is performed with a linear LiCl gradient ranging from 0–0·75M in the same buffer.

III. APPLICATIONS

A. Bacteria

Most investigations have been carried out with *E. coli*. Waller and Harris (1961) were the first to undertake a systematic study of the proteins from the ribosomes of *E. coli*. They found two terminal amino acids after coupling an acid-soluble extract of ribosomes with fluorodinitrobenzene. In starch gel electrophoresis about 20 bands with different mobilities were observed (compare Fig. 2). While these authors envisaged the possibility of an aggregation–disaggregation process occurring between the various proteins, they could find no evidence for any such interaction. Thus the protein bands obtained by electrophoresis on starch gel containing urea retained the same mobility after elution from the gels and upon carrying out further electrophoresis under the same conditions. On carboxymethyl cellulose the proteins could be separated into two fractions with 6M urea containing 0·01N HCl by lowering the pH of the eluting solution from 3·32 to 2·5. After electrophoresis of the two separate fractions, two distinct sets of bands were obtained which together accounted for the total pattern of the 70s proteins (Waller and Harris, 1961; Waller, 1964). A similar phenomenon was also observed after stepwise elution of the ribosomal proteins from carboxymethyl cellulose columns at pH 5·6 by increasing the sodium acetate concentration of the effluent from 0·005M to 0·5M in 6M urea. The elution pattern obtained is shown in Fig. 3. A difference was also obtained between the patterns of the proteins from the two sub-units of the ribosome (Waller, 1964). The subsequent results obtained with polyacrylamide gel electrophoresis (Leboy *et al.*, 1965; Apirion, 1966; Furano, 1966; Traut, 1966; Möller and Chrambach, 1967; Möller and Widdowson, 1967; Gesteland and Staehelin, 1967; Traut *et al.*, 1967; Moore *et al.*, 1968; Krembel and Apirion, 1968) have confirmed Waller's experiments. Thus Traut (1966) used radioactive ribosomal protein to obtain a quantitative estimation of the distribution of ribosomal proteins in the complicated band pattern. He could find no indication of aggregation between protein components, either after re-electrophoresis of the discrete radioactive protein bands with an excess of unlabelled total ribosomal proteins or after mixing the radioactive 30s or 50s sub-unit with the unlabelled 50s or 30s sub-unit, respectively, in the correct proportions and under ionic conditions required to restore the 70s ribosome. The possibility that disulphide-

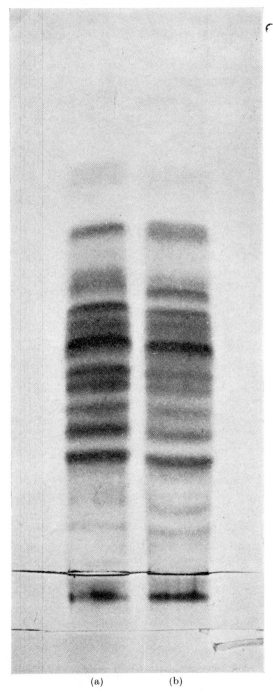

FIG. 2. Starch gel electrophoresis of ribosomal proteins from *E. coli*. (a) Freeze-dried whole ribosomes, (b) acidic acid-soluble protein. Reproduced with permission from J. P. Waller (*J. molec. Biol.* **10**, facing p. 330, 1964).

bridge formation caused some of the numerous bands was also investigated. Conversion of the half-cystine residues present in low amount in the ribosomal proteins (Spahr, 1962) into the S-sulpho cysteine derivative, however, did not simplify the complicated pattern (Waller, 1964). Möller and Chrambach (1967), who compared the pattern of reduced

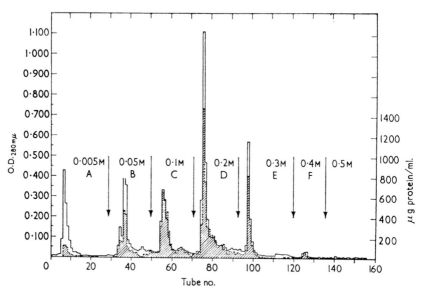

FIG. 3. Fractionation of ribosomal proteins from *E. coli* on a column of carboxymethyl cellulose in 6M urea at pH 5·6 by stepwise elution. Reproduced with permission from J. P. Waller (*J. molec. Biol.* **10**, 328, 1964).

and alkylated ribosomal proteins with the pattern of untreated proteins using polyacrylamide gel electrophoresis, likewise found no indication for intermolecular disulphide bridge formation. Results obtained after incubation of the proteins with β-mercaptoethanol have also failed to show a disulphide interaction (Traut, 1966; Traut *et al.*, 1967). In these experiments the run was performed at pH 4·5 with β-mercaptoethanolamine in the electrophoresis buffers. In contrast to the findings of other authors, we have found a slight increase in the total number of the bands, as well as a marked sharpening of the zones. In our experiments the gels were saturated with β-mercaptoethanol (4% w/v) by a pre-electrophoretic step. The sample was incubated with 6M urea buffer containing 4% (w/v) β-mercaptoethanol and subsequently subjected to electrophoresis again in 6M urea containing mercaptoethanol (compare Fig. 4).

Further evidence for the heterogeneity of the ribosomal proteins from *E. coli* has recently been provided by Traut *et al.* (1967) and by Moore *et al.* (1968). These authors found characteristic and different patterns in fingerprints of tryptic digests from the isolated distinct proteins (see Fig. 5).

B. Mammalian Tissues

Although the ribosomal proteins from animal tissues have not yet been examined as intensively as those from *E. coli*, the data obtained so far indicate that their electrophoretic patterns are also very complex. On starch gel electrophoresis (Terao *et al.*, 1967) or on polyacrylamide gel electrophoresis (Low and Wool, 1967) in 8M urea it is possible to obtain 20–24 distinct bands from the protein isolated from the polysomes of rat liver. By lowering the urea concentration to 4M we were able to show that the number of protein zones increases to about 35 (Bloemendal, 1967; Vennegoor and Bloemendal, unpublished data, Fig. 6). Thirty zones which stained positively with amido-black could be obtained by two-dimensional electrophoresis of rat ribosomal proteins in starch gel containing urea (Terao *et al.*, 1967). In order to rule out the possibility that nascent proteins might account for some of the heterogeneity, treatment of the ribosomes with EDTA may be carried out before extracting the ribosomal proteins. In this case, however, also the patterns show a great number of bands (Terao *et al.*, 1967). Cytoplasmic 50s sub-units and 55s particles of ribosomes isolated from the nucleoli of HeLa cells also yielded numerous protein bands (Warner, 1966; Warner and Soeiro, 1967). Low and Wool (1967) have compared the patterns of ribosomes from rat liver, kidney, skeletal muscle and cardiac muscle and from rabbit reticulocytes by electrophoresis in polyacrylamide gels in 8M urea at pH 4·5 and 8·3. These workers found great similarities in protein patterns. Kedes *et al.* (1966) studied ribosomal proteins from mouse liver and myeloma tumours and also found considerable analogies between the electrophoretic patterns. In our laboratory a comparison has been made between the proteins from rat liver polyribosomes and those from calf lens polyribosomes using electrophoresis at both pH 5·6 in 6M urea and at pH 4·5 in 8M urea. The differences were mainly restricted to intensities in a few bands. The number of proteins isolated from the 30s ribosomal sub-units of lens appeared to be fewer than those from the 50s sub-units. In the latter case the patterns obtained were almost the same as those from total polysomal protein (Fig. 7). At pH 8·9 we could find no bands whatsoever after electrophoresis of the rat liver polysomal proteins,

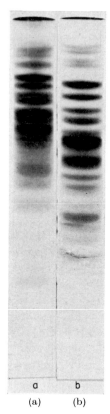

(a)   (b)

Fig. 4. Fractionation of proteins from 70s ribosomes on *E. coli*.

(a) Electrophoresis was carried out on polyacrylamide at pH 5·6 in 6M urea. The preparation and pre-incubation of ribosomes was carried out as described by Nirenberg and Matthei (1961); 70s ribosomes were purified according to Tissières *et al.* (1959). Proteins were extracted from ribosomes using 67% acetic acid (Waller and Harris, 1961). Electrophoresis was carried out as detailed in Section II *D*.

(b) Electrophoresis of the same protein sample as above except that the sample was incubated with 4% (v/v) β-mercaptoethanol. Before applying the sample the gel was electrophoresed with electrode buffers containing 4% (v/v) β-mercaptoethanol. Electrophoresis of the incubated protein sample was then conducted for 154 min at 95 V (6 mA per gel) using renewed electrode buffers containing 4% (v/v) β-mercaptoethanol.

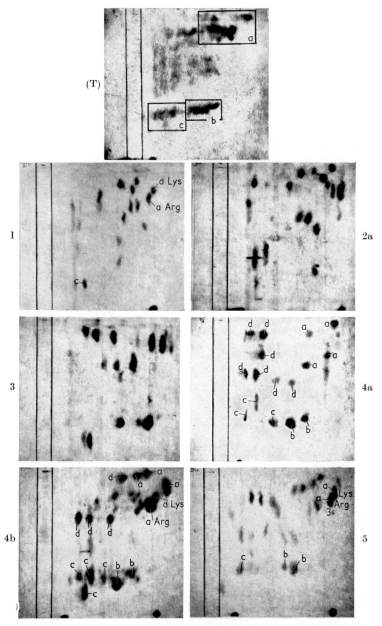

FIG. 5. Fingerprints of total (T) and purified 30s ribosomal proteins (1, 2a, 3, 4a, 4b, 5). Reproduced with permission from R. R. Traut (*J. molec. Biol.*, **31**, 441, 1966).

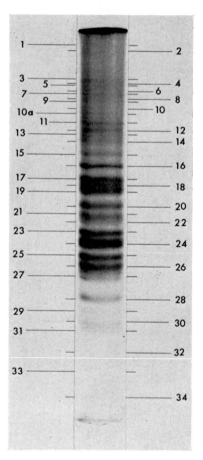

Fig. 6. Electrophoresis pattern of proteins from rat liver polyribosomes. Polyribosomes were prepared according to Bloemendal *et al.* (1967). The polysomal
pellet was suspended in Medium B,* 1·2 ml of the suspension were layered over a
linear sucrose density gradient containing 30% to 10% (w/v) sucrose in Medium B,
and the sample was centrifuged at 25,000 rev/min for 90 min in the SW-25 I rotor
of the Spinco L-II ultracentrifuge. The fractions containing polysomes were
pooled, sedimented by centrifuging for 5½ h at 38,000 rev/min in the Spinco No. 40
rotor and finally resuspended in 1·0 ml of Medium B. Protein was extracted with
67% acetic acid (Waller and Harris, 1961). Electrophoresis was carried out according to Reisfeld *et al.* (1962) except that the gels and electrode buffer contained 4M
urea. The protein sample (1% w/v) was dissolved in the electrode buffer, sucrose
was added to a concentration of 15% (w/v) and the solution layered over the
spacer gel. Electrophoresis was carried out at room temperature for 110 min at
110 V (5 mA per gel).

* Medium B contains 50 mM Tris-HCl, pH 5·6, 25 mM KCl and 10 mM magnesium
acetate.

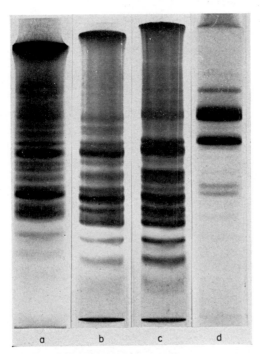

FIG. 7. The protein sample from rat liver polysomes was the same as that used in Fig. 6. Calf lens polyribosomes were prepared according to Bloemendal *et al.* (1966). In order to isolate the 50s and 30s ribosomal sub-units, the polysomes were suspended in a small volume of 0·01M Tris buffer, pH 7·6, containing 0·01M KCl. Na$_2$EDTA was then added to a final concentration of 33 mM. The suspension was layered over a linear sucrose gradient (15–30% w/v sucrose) in 0·01M Tris, pH 7·6, containing 0·01M KCl. The sample was centrifuged at 25,000 rev/min for 16 h in the SW-25 I rotor of the Spinco L-II ultracentrifuge. To the pooled fractions of separated 50s and 30s sub-units was added ethanol at −20° to a final concentration of 66%. Samples were placed overnight in a deep freeze to allow precipitation. Precipitated ribosomal sub-units were sedimented and resuspended in Medium B. Ribosomal proteins were extracted with 67% acetic acid (Waller and Harris, 1961).

Gel electrophoresis patterns shown are from (a) rat liver polysomes, (b) calf lens polysomes, (c) calf lens 50s ribosomal sub-unit; (d) calf lens 30s ribosomal sub-unit. Electrophoresis was performed in polyacrylamide gels containing 8M urea at pH 4·5 as described by Leboy *et al.* (1965). Potassium ferricyanide was used to retard the polymerization during preparation of the gels. The protein sample was layered over the spacer gels as a solution containing sucrose. Electrophoresis was carried out at 160 V for 68·5 min (6 mA per tube).

whereas in the case of the lens polysomal proteins four bands were seen migrating towards the anode (Fig. 8).

Keller *et al.* (1968) have studied the protein moiety of pancreatic ribosomes. Comparison of dog and rabbit ribosomal proteins revealed reproducible species-specific bands in the electrophoretic patterns.

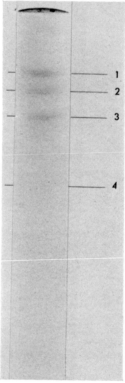

FIG. 8. Pattern of proteins from polysomes of calf lens after electrophoresis in polyacrylamide gels containing 7M urea at pH 8·9 (6 mA per tube for 45 min) according to Bloemendal *et al.* (1962a; 1962b). Bromophenol blue was used to indicate the buffer front. Three hundred μg of the protein samples were applied, three times the amount used in Fig. 7(b).

These zones appeared consistently and independently of the method of preparation of the proteins. Species specificity of embryonic ribosomal proteins has been demonstrated by Mutolo *et al.* (1967).

C. Higher Plants

Cytoplasmic ribosomal proteins from plants have also yielded a very complex electrophoretic distribution pattern on starch gels (Mac-Quillan and Bayley, 1966) or on polyacrylamide gels (Lyttleton, 1968).

While different species in general have different band patterns, closely related species have indistinguishable electropherograms. Ribosomal proteins obtained from different parts of the same plant did not show significant differences. In contrast to this observation, the protein pattern of ribosomes from chloroplasts was plainly different from that of ribosomes from the cytoplasm of the same leaf (Lyttleton, 1968).

D. Other Organisms

Data concerning the characteristics of ribosomal proteins from lower plants and lower animals are still not abundant. However, this class of proteins has already been investigated more thoroughly in the mould *Neurospora crassa* and in the ciliate *Paramecium aurelia*. Ribosomal proteins from those sources are quite similar to those isolated from bacteria, higher plants and mammals. As many as 34 protein zones are observed in the pattern from *Paramecium aurelia* after electrophoresis on polyacrylamide in 8M urea at pH 4·5. The 40s and 25s sub-units, isolated by sucrose density gradient centrifugation at low Mg^{2+} concentration, showed a high degree of dissimilarity (Reisner *et al.*, 1968).

Likewise after gel electrophoresis of ribosomal proteins from *Neurospora crassa*, performed under identical conditions to those above, a very complex pattern was obtained (Alberghina and Suskind, 1967). Furthermore, the amino acid composition of *Neurospora* ribosomal proteins showed a striking similarity to the amino acid composition of ribosomal proteins from other organisms.

Yeast ribosomal proteins have been isolated and characterized by Schmidt and Reid (1968).

E. Ribonuclease Activity

Waller (1964) was able to demonstrate ribonuclease activity in the proteins of the 70s ribosomes and 30s sub-units from *E. coli*. Treatment of the 70s ribosomes in a buffered solution of 0·5M ammonium chloride resulted in removal of the latent ribonuclease (Spirin and Kisselev, 1964). However, it appeared in our hands that even after this treatment a considerable amount of ribonuclease activity could still be demonstrated (Fig. 9). It is generally assumed that the ribonuclease in *E. coli* is bound to the membrane of the cell and becomes attached to the 30s ribosomal sub-unit only after disruption of the cell (Elson and Tal, 1959; Neu and Heppel, 1964; Waller, 1964). It is apparent then that the enzyme once adsorbed cannot be removed completely from the ribosome by salt treatment. This is in agreement with the findings of Schell (1966).

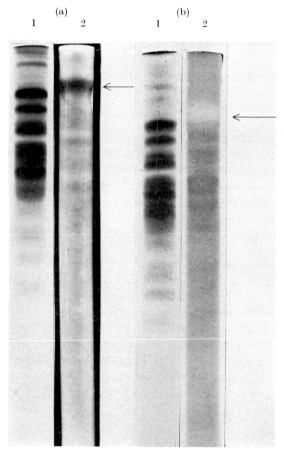

Fig. 9. Ribonuclease activity in proteins of 70s ribosomes from *E. coli* B. Ribosomes were prepared and preincubated according to Nirenberg and Matthaei (1961). The ribosomal suspension was layered over a buffer consisting of 10% (w/v) sucrose, 0·5M NH$_4$Cl, 0·10M magnesium acetate, 0·01M Tris-HCl and 0·06M β-mercaptoethanol, pH 7·4. Centrifugation was carried out for 5 h at 25,000 rev/min in the SW-25 I rotor of the Spinco L-50 ultracentrifuge. The colourless pellet was suspended in the standard buffer of Matthaei and Nirenberg (1961). Protein was extracted with 67% acetic acid. Electrophoresis was carried out on polyacrylamide at pH 5·6 in 6M urea (see Section II *D*). Ribonuclease activity was detected as described in Section II *E*. (*a*)*1* and (*b*)*1* are halves of gels stained for protein. (*a*)*2* is one half of a gel incubated with 1% (w/v) RNA in distilled water after which the RNA was precipitated with 0·1N HCl. With gel (*b*)*2* the incubation was carried out in 0·1% (w/v) RNA in 0·1M sodium acetate, pH 5·0. This gel was then stained with acridine orange. One hundred μg of protein were applied to each gel. Electrophoresis was carried out at 80 V for 150 min (6 mA per tube). Ribonuclease activity is indicated by the arrows.

It is also worth noting that chromatography of the ribosomes on DEAE cellulose has shown ribonuclease to be much more strongly attached to the ribosomes than is deoxyribonuclease (Furano, 1966).

The ribonuclease activity of the protein isolated from rat liver polysomes appears to be very low when compared to the activity in ribosomal proteins from *E. coli* (Fig. 10). In the proteins of calf lens ribosomes we were not able to demonstrate any ribonuclease activity whatsoever with the present technique (Bloemendal *et al.*, 1969).

IV. CORRELATION OF ELECTROPHORETIC CHARACTERIZATION WITH OTHER DATA

It is extremely difficult at the present time to state with accuracy the exact number of protein components in the protein moiety of the ribosome. In 1962 Spitnik-Elson had already concluded from precipitation experiments with ribosomal protein from *E. coli* that these proteins are a heterogeneous mixture with different isoelectric points. Originally Waller and Harris (1961) and Waller (1964), using starch gel electrophoresis as a criterion, claimed the presence of 20 components. This number increased to 24–25 using chromatography on carboxymethyl cellulose together with electrophoresis, or when the isolated 50s and 30s sub-units were examined by electrophoresis alone. Electrophoresis on polyacrylamide at pH 4·5 yielded 28 different protein zones detectable by staining with amido-black (Leboy *et al.*, 1965; Furano, 1966). Using proteins labelled with [^{35}S]cysteine together with [^{35}S]methionine, even more zones could be demonstrated. At the present time 19 components are found in the 50s sub-unit and 11–13 components in the 30s sub-unit using polyacrylamide gel electrophoresis (Traut *et al.*, 1967; Moore *et al.*, 1968).

On the basis of amino terminal analyses an average molecular weight of 25,000 was calculated by Waller and Harris (1961) for the proteins of the 70s ribosome. This number is close to the values obtained by Möller and Chrambach (1967) who found on the basis of sedimentation diffusion experiments an average molecular weight of 26,000. On the basis of sedimentation equilibrium studies, a weight-average molecular weight of 23,000 and a z-average molecular weight of 27,000 were calculated. A 70s ribosome, the protein part of which corresponds to about 10^6 daltons, would then be composed of 40 protein sub-units, 26 of which would occur in the 50s sub-unit and about 14 in the 30s sub-unit (Waller, 1964). Moore *et al.* (1968), however, determined the weight-average molecular weights of total 30s and 50s proteins to be 14,000 and 15,200, respectively. In the same solvent, however, the total 70s protein had an apparent average molecular weight of 21,000, a value in reasonable agreement with those of the authors quoted above. After mixing the 30s and 50s sub-units, an average molecular weight of 21,000 was again obtained. Moore *et al.* (1968) conclude, therefore, that the value of 21,000 is too high and is probably caused by a strong interaction between the 30s and 50s proteins.

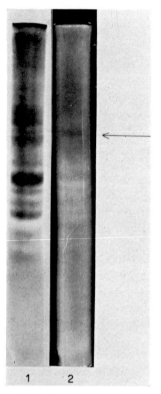

Fɪɢ. 10. Ribonuclease activity in proteins from polysomes of rat liver. Conditions were as in Fig. 9(a). Four hundred μg of protein were applied. A small area of faint ribonuclease activity is indicated by the arrow.

The number of proteins in the 70s ribosome, therefore, may be greater than the figure of 40 quoted above. Actually 20 different protein species could be distinguished in the 30s sub-unit by comparing the electrophoretic mobilities of the total ribosomal proteins and the purified 30s proteins in gels containing 0·2% and 0·8% bisacrylamide. By gel filtration on Sephadex G-100 in urea and by chromatography on carboxymethyl cellulose in urea, Moore *et al.* (1968) demonstrated that some of the 13 protein zones obtained by electrophoresis on acrylamide gels (0·2% bisacrylamide) consisted of different proteins. From these results and from the data obtained with sedimentation equilibrium studies the authors assumed that 23 proteins must exist in the 30s sub-unit and 43 proteins in the 50s sub-unit.

The possibility that aggregation causes heterogeneity of ribosomal proteins seems definitely to be ruled out at the present time. Waller (1964) re-electrophoresed the protein zones obtained on starch gels and observed that each band retained the same mobility. Traut (1966) confirmed this finding using radioactively labelled proteins. Disulphide-bridge formation does not contribute to the heterogeneity (Waller, 1964; Traut, 1966; Möller and Chrambach, 1967) nor does a combination of phosphate or nucleotide binding, partial deamidation or carbamylation (Möller and Chrambach, 1967). On the other hand, our experiments (compare Fig. 4) indicate a slight increase in the number of zones after adding -SH reagents to the protein sample and to the gel. Also in the case of *Neurospora crassa* reduction of the extracted ribosomal protein with β-mercaptoethanol, with or without subsequent alkylation with iodoacetamide, resulted in the changes in the electrophoretic profile.

From fingerprinting studies of purified proteins from *E. coli* labelled with sulphur-35 it is concluded that at least 19 proteins with a molecular weight of 25,000 must have a different primary structure (Traut *et al.*, 1967). Thirteen proteins, all from the 30s sub-unit, have already been purified and examined. In tryptic fingerprints the number of spots and their location on peptide maps appeared to be a distinctive characteristic of each protein (Moore *et al.*, 1968). One or two of these proteins do not possess a free amino terminal group (Möller and Castleman, 1967; Traut *et al.*, 1967; Moore *et al.*, 1968).

Attention has also been paid to the problem of the growing peptide chains on the isolated ribosomes and of the adsorption of non-specific proteins as factors contributing to the heterogeneity of ribosomal proteins. The analyses of the NH_2-terminal amino acids of total ribosomal protein by Waller and Harris (1961) and Waller (1964) and of the protein fractions obtained after separation on carboxymethyl cellulose, however, demonstrate that, with the exception of one fraction, methionine and alanine are always the major amino terminals. Moreover, the electrophoretic pattern changed only slightly after chromatography of the ribosomes on DEAE cellulose (Furano, 1966). Furthermore, the 40s ribonucleoprotein particle which is a ribosomal precursor lacked only four of the proteins contained in the 50s sub-unit as shown by chromatography on carboxymethyl cellulose in sodium acetate buffer, containing 6M urea (Otaka *et al.*, 1967). The presence of nascent protein or contamination with cytoplasmic proteins, therefore, would not seem to be serious causes of heterogeneity. Hosokawa *et al.* (1966) and Gesteland and Staehelin (1967) showed that a more or less specific separation of proteins from the 30s and 50s sub-units may be obtained by density gradient centrifugation in CsCl, which results in the separation of 23s particles and free proteins in the case of the 30s particles, and 42s particles and free proteins in the case of the 50s particles. The proteins of the 23s particles lacked some bands found in the 30s pattern and the proteins of the 42s particles likewise lacked some proteins present in the 50s pattern. Mixing 42s and 23s particles cannot restore the *in vitro* protein synthetic activity of the ribosome. Active ribosomes however, may be obtained after mixing the 42s particles and their free proteins with 23s particles and their free proteins. It appeared that the ability of the 30s sub-unit to bind messenger RNA had been lost in the case of the 23s particle and could be restored only after mixing the 23s particle with the proteins freed from the 30s particle. Similarly reconstituted 50s particles regained their capacity to bind transfer RNA after mixing the 42s particles with the proteins freed from the 50s particles (Raskas and Staehelin, 1967). Recent results of Friedman *et al.* (1968) are in agreement with this finding. These authors were able to demonstrate that the proteins stripped from 50s sub-units with the aid of CsCl could be used to obtain a specific antiserum. This antiserum reacted with 50s and 70s ribosomes while core particles (40s and 23s) were non-reactive. Ribosomes preincubated with transfer RNA exhibited a significant reduction in serological reactivity. However, ribosomes preincubated with transfer RNA preparations which were rendered incapable of binding to the ribosome by removing the terminal nucleoside showed the usual serological reactivity. These results demonstrate that the protein determinant studied which was split from the 50s ribosome was located at or near to the binding site for transfer RNA. Hence it is tempting to believe that the proteins stripped from the ribosome by treatment with CsCl represent components essential for the function of the ribosome.

As noted with *E. coli* ribosomal proteins (Gesteland and Staehelin, 1967), the "split protein" from *Neurospora* prepared by treatment of the ribosomes with CsCl showed acidic and alkaline electrophoretic profiles quite distinct from that of the total ribosomal protein. A further indication that the latter proteins constitute a distinct part of the ribosome was given by comparing the antigenic properties of total ribosomal proteins with "split protein" using immunodiffusion methods. The "split protein" was homologous with only one of the components of the total ribosomal protein (Alberghina and Suskind, 1967).

Taking all the evidence together it may be stated that the ribosomal protein of *E. coli* consists of perhaps 66 protein components of which 13 are certainly different. The main amino terminal groups are alanine, methionine, serine and threonine. In bacteria there seems to exist a species specificity with regard to ribosomal proteins since, at least in the case of *B. cereus, B. megaterium* and *Micrococcus lysodeiktus*, different electrophoretic patterns are obtained (Waller, 1964). Also by chromatography on carboxymethyl cellulose of sub-units of *E. coli* ribosomes labelled with tritium and sub-units of ribosomes from *Sarcina lutea, B. cereus, B. subtilis* or *B. megaterium* labelled with carbon-14, no apparent similarities were found in the elution patterns (Otaka *et al.*, 1968). The electrophoresis patterns of different strains of *E. coli* are very similar. Recently, however, some differences have been found in the patterns of the ribosomal proteins from strains K_{12} and Q_{13} (Q_{13} is derived from a K strain) and from *lir* mutants of *E. coli* (Leboy *et al.*, 1965; Otaka *et al.*, 1968; Krembel and Apirion, 1968). However, the ribosomal protein of *Neurospora crassa* seems to possess some antigenic determinant present in *E. coli* protein (Reisner *et al.*, 1968).

Studies by Nakada (1968) revealed that proteins of so-called relaxed particles (formed by a methionine-requiring, relaxed mutant of *E. coli*) are, for the major part, derived from ribosomal proteins. This is a strong indication that pre-existing ribosomal protein can combine with newly synthetized ribosomal RNA.

The electrophoretic pattern of mammalian ribosomal proteins is strikingly different from the pattern of *E. coli* ribosomal proteins (compare Fig. 11 with Fig. 7). The differences are still greater when the patterns of the 30s and 50s sub-units of *E. coli* ribosomes (Traut, 1966) are compared with the subunits of, for instance, plasma cell tumour (Kedes *et al.*, 1966) or calf lens ribosomes (Fig. 7). It appears that in mammalian tissues relatively few of the zones obtained after acrylamide gel electrophoresis of 30s proteins are present in the pattern from the 50s proteins.

In studying ribosomal proteins of mammals much attention has been directed to the possibility of contamination of ribosomal proteins with cytoplasmic proteins occurring during isolation of the ribosomes. It does seem that purified HeLa cell ribosomes are able to adsorb non-specific proteins (Warner and Pène, 1966). These proteins can be removed from the ribosomes by treatment with a moderately high salt concentration in the presence of EDTA (Warner, 1966; Warner and Pène, 1966). However, the possibility of losses of true ribosomal proteins by this procedure is not completely ruled out.

Homology seems to exist between the nascent ribosomal particles and the ribosomal sub-units. Thus in plasma cell tumours the electrophoretic pattern of the 45s post-ribosomal particles is homologous to the 30s pattern and the pattern of the 60s post-ribosomal particles resembles the 50s pattern (Kedes *et al.*, 1966). Also in HeLa cells the pattern of the 55s particles is similar to the pattern of the 50s sub-units (Warner,

FIG. 11. Pattern of proteins from 70s ribosomes of *E. coli*. The preparation of the ribosomes and the extraction of the proteins were as described in Fig. 9. Electrophoresis was performed in polyacrylamide at pH 4·5 in 8M urea.

1966; Warner and Soeiro, 1967). From studies using protein labelled with carbon-14 and tritium, Warner (1966) concluded that the different fractions obtained after electrophoresis on polyacrylamide gels must represent distinct species of polypeptide chains.

Attention has also been focused on the metabolic activity of ribosomal proteins in HeLa cells (Warner, 1966) and in rat liver (Terao *et al.*, 1967). In HeLa cells Warner (1966) found that the cytoplasm contains pools of ribosomal proteins which allow maturation of ribosomes in the presence of cycloheximide. He distinguished three classes of protein, termed A, B and C depending on the characteristics of their association with the ribosomal RNA. The A proteins appeared to enter the cytoplasmic ribosomes only in the company of newly formed ribosomal RNA. The B proteins exchanged with the soluble proteins *in vivo* but were firmly bound to the ribosomes after extraction from the cell. The C proteins represent the proteins less firmly bound to the ribosomes (Warner, 1966; Warner and Soeiro, 1967). The latter proteins are in equilibrium with the soluble proteins but they can be reversibly dissociated from the ribosomes at an ionic strength of 0·2 (Warner and Soeiro, 1967). The A class proteins seemed to appear at different rates among the various proteins of the 50s sub-units (Warner, 1966).

Heterogeneity in the metabolic activity of ribosomal proteins has also been demonstrated by Terao *et al.* (1967). The less basic proteins were metabolically more active than the more basic ones. The half-life of ribosomal proteins as a whole appeared to be 90–80 h, a value comparable with that of the total proteins of other cell components such as mitochondria, nuclei and cell sap.

In general, the electrophoretic patterns of ribosomal proteins from mammalian tissue exhibit great similarities, not only with respect to the proteins from different tissues of the same animal, but also with respect to proteins from different animals. However, a difference was observed by Cohn (1967) who compared the proteins from rat liver and rabbit reticulocytes by electrophoresis in starch gel containing urea in 0·02N HCl. Although differences were observed in the patterns, he found the same main amino terminal residues, viz. alanine, glycine, proline and serine, in proteins from both sources. Also only small differences in the proportions of these end groups and of the total amino acid composition were found in the proteins from both species.

The molecular weights of mammalian ribosomal proteins lie probably in about the same range as those found in *E. coli*. From amino terminal analysis in rat liver ribosomes Cohn and Simson (1963) and Cohn (1967) calculated values ranging from 24,000 to 40,000.

In conclusion we may say that although a growing number of ribosomal proteins are being chemically characterized, much more work has yet to be done to establish both their exact roles in the function of the ribosomes, and also the precise manner in which they are integrated into the structure of the ribosome. In this connection a statement by Traub *et al.* (1967) should be recalled. 'Chemically distinct (ribosomal) proteins also are functionally distinct'.

REFERENCES

Alberghina, F. A. M. and Suskind, S. R. (1967). *J. Bact.* **94**, 630.
Apirion, D. (1966). *J. molec. Biol.* **16**, 285.
Arnstein, H. R. V., Cox, R. A. and Hunt, J. A. (1964). *Biochem. J.* **92**, 648.

Bloemendal, H. (1967). *In*, "Electrophoresis" (Milan Bier, ed.) Vol. 2, Academic Press, New York and London.

Bloemendal, H., Bont, W. S., Jongkind, J. F. and Wisse, J. H. (1962a). *Expl Eye Res.* **1**, 300.

Bloemendal, H., Jongkind, J. F. and Wisse, J. H. (1962b). *Chem. Weekbl.* **58**, 501.

Bloemendal, H., Bont, W. S. and Benedetti, E. L. (1964). *Biochim. biophys. Acta* **87**, 177.

Bloemendal, H., Schoenmakers, J. G. G., Zweers, A. Matze, R. and Benedetti, E. L. (1966). *Biochim. biophys. Acta* **123**, 217.

Bloemendal, H., Bont, W. S., de Vries, M. and Benedetti, E. L. (1967). *Biochem. J.* **103**, 177.

Bloemendal, H., Vennegoor, C. and Konings, R. (1969). *Expl Eye Res.* **8**, 220.

Bretscher, M. S. and Jones, O. W. (1967). *In*, "Techniques in Protein Biosynthesis" (P. N. Campbell and J. R. Sargent, eds), Vol. 1, p. 217, Academic Press, London and New York.

Cohn, P. (1967). *Biochem. J.* **102**, 735.

Cohn, P. and Simson, P. (1963). *Biochem. J.* **88**, 206.

Cox, R. A. and Arnstein, H. R. V. (1962). *Biochem. J.* **89**, 574.

Curry, J. B. and Hersh, R. T. (1966). *Biochim. biophys. Acta* **129**, 435.

von der Decken, A. (1967). *In*, "Techniques in Protein Synthesis" (P. N. Campbell and J. R. Sargent, eds), Vol. 1, p. 65, Academic Press, London and New York.

Elson, D. and Tal, M. (1959). *Biochim. biophys. Acta* **36**, 281.

Florini, J. R. and Breuer, C. B. (1966). *Biochemistry, N.Y.* **5**, 1870.

Friedman, D. I., Olenick, J. G. and Hahn, F. E. (1968). *J. molec. Biol.* **32**, 579.

Furano, A. (1966). *J. biol. Chem.* **241**, 2237.

Gesteland, R. F. and Staehelin, T. (1967). *J. molec. Biol.* **24**, 149.

Hosokawa, K., Fujimura, R. K. and Nomura, M. (1966). *Proc. natn. Acad. Sci. U.S.A.* **55**, 198.

Kedes, L. H., Koegel, R. J. and Kuff, E. L. (1966). *J. molec. Biol.* **22**, 359.

Keller, P. J., Cohen, E. and Hollinshead Beeley, J. A. (1968). *J. biol. Chem.* **243**, 1271.

Korner, A. (1961). *Biochem. J.* **81**, 168.

Krembel, J. and Apirion, D. (1968). *J. molec. Biol.* **33**, 363.

Leboy, P. S., Cox, E. C. and Flaks, J. G. (1965). *Proc. natn. Acad. Sci. U.S.A.* **52**, 1367.

Low, R. B. and Wool, I. G. (1967). *Science, N.Y.* **155**, 330.

Lyttleton, J. W. (1968). *Biochim. biophys. Acta* **154**, 145.

MacQuillan, A. M. and Bayley, S. T. (1966). *Can. J. Biochem.* **44**, 1221.

Master, R. W. P. and Srinivasa Rao, S. (1961). *J. biol. Chem.* **236**, 1986.

Matthaei, J. H. and Nirenberg, M. W. (1961). *Proc. natn. Acad. Sci. U.S.A.* **47**, 1580.

Möller, W. and Castleman, H. (1967). *Nature, Lond.* **215**, 1293.

Möller, W. and Chrambach, A. (1967). *J. molec. Biol.* **23**, 377.

Möller, W. and Widdowson, J. (1967). *J. molec. Biol.* **24**, 367.

Moore, P. B., Traut, R. R., Noller, H., Pearson, P. and Delius, H. (1968). *J. molec. Biol.* **31**, 441.

Mutolo, V., Gindice, G., Hopps, V. and Donatuti, G. (1967). *Biochim. biophys. Acta* **138**, 214.

Nakada, D. (1968), *J. molec. Biol.* **29**, 473.

Neu, H. C. and Heppel, L. A. (1964). *Biochem. biophys. Res. Commun.* **14**, 109.
Nirenberg, M. W. and Matthaei, J. H. (1961). *Proc. natn. Acad. Sci. U.S.A.* **47**, 1588.
Otaka, E., Itoh, T. and Osawa, S. (1967). *Science, N.Y.* **157**, 1452.
Otaka, E., Itoh, T. and Osawa, S. (1968). *J. molec. Biol.* **33**, 93.
Rampersad, O. R. and Wool, I. G. (1965). *Science, N.Y.* **149**, 1102.
Raskas, H. J. and Staehelin, T. (1967). *J. molec. Biol.* **23**, 89.
Reisfeld, R. A., Lewis, U. J. and Williams, D. E. (1962). *Nature, Lond.* **195**, 281.
Reisner, A. H. Rowe, J. and MacIndoe, H. M. (1968). *J. molec. Biol.* **32**, 587.
Rendi, R. and Hultin, T. (1960). *Expl Cell Res.* **19**, 253.
Richards, E. G., Coll, J. A. and Gratzer, W. B. (1965). *Analyt. Biochem.* **12**, 452.
Schell, P. L. (1966). *Nature, Lond.* **210**, 1157.
Schmidt, J. and Reid, B. R. (1968). *Biochem. biophys. Res. Commun.* **31**, 654.
Schoenmakers, J. G. G., Zweers, A. and Bloemendal, H. (1967). *Biochim. biophys. Acta* **145**, 120.
Schweet, R. S., Lamfrom, H. and Allen, E. (1958). *Proc. natn. Acad. Sci. U.S.A.* **44**, 1029.
Smithies, O. (1955). *Biochem. J.* **61**, 629.
Sober, H. A. and Peterson, E. A. (1962). *In*, "Methods in Enzymology" (S. P. Colowick and N. O. Kaplan, eds) Vol. 5, Academic Press, New York and London.
Spahr, P. F. (1962). *J. molec. Biol.* **4**, 395.
Spirin, H. S. and Kisselev, N. A. (1964). *Abstracts of VIth Int. Congr. Biochem. New York, IUB,* **32**, 32.
Spitnik-Elson, P. (1962). *Biochim. biophys. Acta* **61**, 625.
Spitnik-Elson, P. (1965). *Biochem. biophys. Res. Commun.* **18**, 557.
Summers, D. F., Maizel, J. V. and Darnell, J. E. (1965). *Proc. natn. Acad. Sci. U.S.A.* **54**, 505.
Terao, K., Katsumi, H., Sugano, H. and Ogata, K. (1967). *Biochim. biophys. Acta* **138**, 369.
Tissières, A., Watson, J. D., Schlessinger, D. and Hollingworth, B. R. (1959). *J. molec. Biol.* **1**, 221.
Traub, P., Hosokawa, K., Craven, G. R. and Nomura, M. (1967). *Proc. natn. Acad. Sci. U.S.A.* **58**, 2430.
Traut, R. R. (1966). *J. molec. Biol.* **21**, 571.
Traut, R. R., Moore, P. B., Delius, H., Noller, H. and Tissières, A. (1967). *Proc. natn. Acad. Sci. U.S.A.* **57**, 1294.
Waller, J. P. (1964). *J. molec. Biol.* **10**, 319.
Waller, J. P. and Harris, J. I. (1961). *Proc. natn. Acad. Sci. U.S.A.* **47**, 18.
Warner, J. R. (1966). *J. molec. Biol.* **19**, 383.
Warner, J. R. and Pène, M. G. (1966). *Biochim. biophys. Acta* **129**, 359.
Warner, J. R. and Soeiro, R. (1967). *Proc. natn. Acad. Sci. U.S.A.* **58**, 1984.
Wettstein, F. O., Staehelin, T. and Noll, H. (1963). *Nature, Lond.* **197**, 430.

CHAPTER 5

Animal Operative Techniques
(In the Mouse, Rat, Guinea Pig and Rabbit)

H. B. Waynforth

Courtauld Institute of Biochemistry
Middlesex Hospital Medical School, London, England

I. INTRODUCTION

Tissues and tissue products form the basic materials for the research work of the biochemist. It follows that some knowledge of the biology of the laboratory animal is a prerequisite in the successful performance of this research. Until quite recently, extensive "preparation" of the animal prior to chemical investigation of its components was rarely carried out. The considerable progress made in modern biochemistry involving refinements both in chemical technique and in instrument technology has made possible, and indeed essential, a more complete integration of animal experimentation with chemical analysis.

The literature abounds with information on the experimental manipulation of animals but there are very few instances in which the various basic techniques have been assembled to serve most of the needs of the investigator and especially the person interested in protein biosynthesis. The present chapter has been designed to do just this in an endeavour to obviate the need to spend time and effort in gathering information from widespread references. A further purpose of the chapter is to make known to the reader the extent of the techniques available. Some selection inevitably has been made so that only the most useful techniques are described.

Each method has been given in some detail so that the investigator should be able to attempt the procedures without further advice. It cannot be too strongly emphasized, however, that in work with animals, there is no substitute for familiarity through practical experience. The realization of this ideal is not directly possible in the British Isles, since *under the Cruelty to Animals Act of 1876 live animals cannot be used solely for the purpose of attaining manual skill.* A useful degree of dexterity and confidence, however, can be gained by the use of freshly killed animals.

II. INJECTION TECHNIQUES
A. General Information
I. Handling of Animals

Injecting materials into animals is a simple procedure providing the animals are held firmly and correctly. Proper and frequent handling causes animals to become docile and easier to work with. With such animals there is less chance of the investigator being scratched or bitten and consequently becoming agitated. A nervous investigator often transmits his nervousness to the animal he is handling so that all animals should be approached slowly and steadily and at least with an outward show of confidence! To attempt to catch small laboratory animals from behind in a somewhat indecisive fashion is extremely ill-advised.

Mice and rats can be picked up by their tails, preferably near the base. Rats, however, can shed the skin of their tails and should be gripped about the body if it is intended to hold them for a lengthy period. For the purpose of injection, mice are held as in Fig. 1. The tail is placed between the third and fourth fingers and the animal is held by the scruff of the neck with the thumb and index finger. The neck should be held so that head movement is minimized. With practice, a fold can be formed with the free second finger from the skin of the back, into which a subcutaneous injection can easily be made.

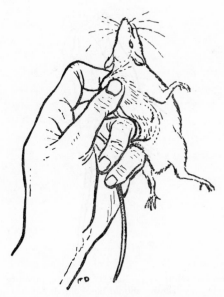

FIG. 1. Method of holding the mouse for injection. A fold can be produced from the loose skin of the back to facilitate subcutaneous injection.

Two methods of holding rats can be used depending on the type of injection to be made. The more usual way is to grip the animal with the flat of the hand on its back, the thumb passing in front of or behind one forelimb and ending well up under the lower jaw (Fig. 2). The fingers should be placed comfortably around the body and should extend as near as possible into the region of the groin. Again there should be a minimum of head movement. The second method is particularly useful for intraoral administration when it is essential to keep the head in line with the body and to keep the mouth open (Fig. 3). The animal is held by the loose skin of the neck and back, with the skin around the neck being pulled tight but not, of course, to the extent of causing excessive interference with respiration. The head should be immobilized almost completely if the procedure is executed correctly. Infantile rats can best be held by the same method as is used for mice.

9

Guinea pigs are picked up and held as in Fig. 2 for rats. Some support of the rear should be given with the other hand to large animals or those in an advanced stage of pregnancy.

Rabbits can be picked up most conveniently by the scruff of the neck. Support of the rear should be given either with the other hand or by

FIG. 2. Holding the rat. Method 1. Note that undue pressure on the throat and thorax should be avoided.

placing the animal under the opposite arm. The animal must not be allowed to kick as it can damage both itself and the person holding it. The mode of holding rabbits for injection is discussed under the various routes for injection (see Figs. 4, 8, 11, 13).

2. Solutions for Injection

A number of questions often arise concerning the suitability of solutions for injection into the animal: "what volume can be used?", "will the pH affect the animal adversely?" and so forth. Unfortunately precise answers are not forthcoming, principally because of the considerable variation to be found in responses between different species of animals and very often even between animals of the same species. However, generalizations may be made which can act as a guide for the unwary operator and which usually suffice to overcome satisfactorily most problems that arise.

A. Solvents. Many substances can conveniently be injected in distilled water or physiological saline (0·85% NaCl), both of which are ideal vehicles physiologically. For reasons of solubility or rate of absorption, however, some substances may require a more complex solvent to render them suitable for administration. One or more of the

Fig. 3. Holding the rat. Method 2.

following materials have at various times been combined with distilled water or physiological saline to provide a suitable injection medium: 60% (v/v) propane-1:2-diol; 0·5% (w/v) carboxymethyl cellulose; 10% (v/v) Tween 80; 10% (v/v) ethanol and 50% (v/v) dimethylformamide. All these vehicles can be administered by any of the injection routes available but the concentrations mentioned are the maximum practicable and in many cases it is possible, and indeed desirable, that a lower concentration be used.

Vegetable oils are suitable for injection when lipid-soluble substances are to be administered but they cannot be injected via the intravenous route. If it is necessary to inject such lipoidal substances directly into the circulation then it may be possible to do so as a fine aqueous suspension or in a 15% (v/v) oil–water emulsion using, for example, lecithin as the emulsifier. It should be noted, however, that some animals, especially guinea pigs, are adversely sensitive to intravenous injections of suspended particles.

B. Quantity. The maximum quantity of a solution that can be injected into an animal will depend mainly on the route of administration. A useful guide is to use a volume of up to 10 ml/kg of body weight.

In some instances a greater volume can be administered if several injection sites are utilized. For intramuscular injection it is doubtful whether more than about 1·0 ml for the rabbit and 0·25 ml for the adult rat will be retained at the site of injection. It is possible to administer as much as 50 ml/kg of body weight into the stomach providing that the solution is approximately isotonic. Large quantities can also be introduced into the peritoneal cavity.

C. *pH*. Animals can tolerate injections over a fairly wide range of pH. For all routes of administration, a working range is in the region of pH 4·5–8·0. Solutions of greater acidity can be tolerated in the stomach if the titratable acidity does not exceed the equivalent of 0·1N HCl but alkaline solutions are not well accepted by the stomach. The widest tolerance to pH is shown by the intravenous route because of the strong buffering capacity of the blood, followed by the intramuscular and then the subcutaneous routes.

D. *Absorption*. The rate of absorption of substances after injection depends most often on the route of administration and the vehicle used. Complete absorption from oil or viscous solutions is very much slower than from water. Absorption is relatively rapid after intravenous administration and is 2–4 times faster than the rate of absorption from an intraperitoneal injection. The rate of absorption from the subcutaneous and oral routes is variable but is usually less rapid than from the intraperitoneal route. The slowest rate of absorption is from intramuscular injection and in some instances absorption may be complete only after several months.

E. *Rate of injection*. The rate at which solutions can be injected intravenously rests mainly on the concentration of the material in solution, its physiological effect and the pH of the solution. Physiological saline for example, can be injected at a rate of 3 ml/min or more, depending on the size of the animal, but in general an intravenous injection should be given slowly, paying attention to the state of the animal during the injection.

B. Routes of Administration

I. Subcutaneous

The injection is usually made under the skin of the back and sides but any other area can be used if these sites are unavailable. After passing the needle through the skin, the syringe should be moved about gently to reveal the whereabouts of the tip of the needle and thus ascertain that the needle is truly subcutaneous. A successful subcutaneous injection most often results in the formation of a bleb during the discharge of solution from the syringe. Ideally the whole of the needle shaft should be subcutaneous as this ensures a minimum of leak-back of solution. Irrespective of the size of the animal, the smallest size of hypodermic needle should be used compatible with the material being

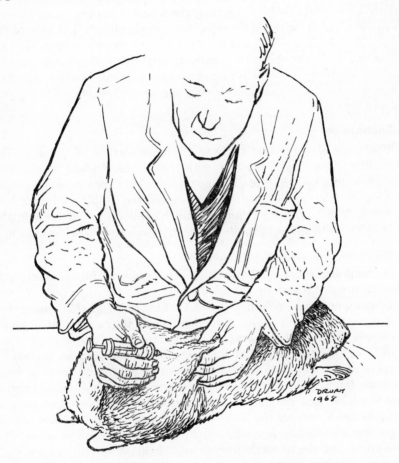

FIG. 4. Restraining the rabbit for subcutaneous injection. This method can also be used when giving an intramuscular injection.

injected. A 26 gauge* needle, for instance, is suitable for non-viscous aqueous solutions, whereas a 23 or 21 gauge needle is used for viscous solutions or for suspensions.

Rabbits can be injected with the minimum of restraint, the injection being made either while the animal is in its cage or while held on a table, close to the body (Fig. 4).

* Standard wire gauge.

2. *Intraperitoneal*

The injection is made into the lower lateral part of the abdomen but not as far posterior as the groin. The tip of the needle should first be inserted subcutaneously and a final short thrust is then made through the abdominal muscles. If the injection is made into the gut by mistake, fluid will often be seen issuing from the rectum during or soon after the injection. It is necessary to insert only the tip of the needle into the peritoneal cavity. Guinea pigs and rabbits are most conveniently held by an assistant who can then present the ventral surface of the animal for the injection.

3. *Intramuscular*

The usual site for this route is the musculature of the rear limb assembly. The *biceps femoris* and the *semitendinosus* and *gluteus maximus* muscles (thigh and rump, respectively) are most frequently employed. The needle must not be inserted too deeply otherwise bone is encountered. The injection procedure is simplified if the animal is held by an assistant but rabbits can be injected easily if held as in Fig. 4.

4. *Intravenous*

A. *Sublingual vein*. (M, Rt, Gp)*. There is little information in the literature on the use of the sublingual veins for injection in small laboratory animals (Waynforth and Parkin, 1969). Yet it would seem to the author that injection into these veins is the method of choice since it is simple, quick, involves no surgery or prior preparation, rarely results in the production of an haematoma and can be repeated a number of times at close intervals. The sublingual veins are to be found on both lateral aspects of the ventral surface of the tongue (Fig. 5). For the injection, the animal is anaesthetized and placed on its back with its head towards the investigator. A gauze pad is placed over its nose and anaesthesia is continued by drop application of ether. The tongue is pulled out with forceps and held at its tip, under slight tension, with the thumb and index finger, in such a way that it presents a slight convex curvature and the flow of blood in the vein to be injected is not obstructed. The veins are exceedingly superficial and it is necessary to enter them almost horizontally to the surface of the tongue. Entry into any vein can usually be judged to be successful if the needle slides freely backwards and forwards in the vein and if the tip of the needle can be seen when raised and cannot be seen when lowered, i.e. reflecting blood passing under and over the needle tip, respectively. In large veins it should be possible to withdraw a little

* M, mouse; Rt, rat; Gp, guinea pig; Rb, rabbit.

blood into the syringe if the blood vessel has been entered correctly. After the injection, the site of entry is immediately covered by a small cotton-wool pledget which, together with the tongue, is pushed into the mouth effectively acting as a haemostat (Fig. 6). On recovery from the anaesthetic, the animal spits out the pledget and bleeding rarely occurs.

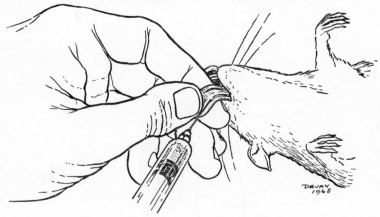

Fig. 5. Sublingual vein injection in the rat.

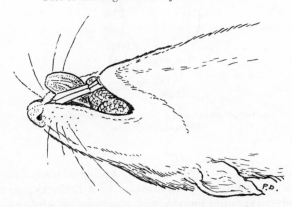

Fig. 6. Cotton-wool pledget placed between the tongue and lower jaw to prevent bleeding after injection into the sublingual veins.

A possible disadvantage in the method is the relatively rapid rate of recovery of the animal during the injection, since air and not ether vapour is being breathed in through the mouth. This becomes a problem only in the case of a lengthy injection, but it can be overcome to a large extent by plugging the mouth with a cotton-wool pledget which is removed after the injection. Depending on the size of the animal, a 26 or 30 gauge needle can be used.

The method is particularly suitable for the infantile and adult rat and can be used with advantage in the mouse. The injection is possible in the guinea pig but only in completely anaesthetized and relaxed animals.

B. Ear vein. (Gp, Rb). This is the method of choice in the rabbit. The large vein is situated at the lateral outer aspect of the ear (Fig. 7). The hair is clipped short at the site of injection and the vein is dilated

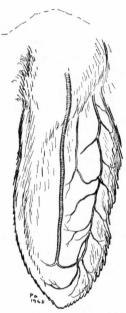

Fɪɢ. 7. Ear of the rabbit showing the laterally placed vein and the central artery (semi-diagrammatic).

either by holding the ear close to a lamp bulb and massaging the ear in the region of the vein or by application of a little xylene or toluene. The latter solvents should be removed, however, by wiping with ethanol prior to making the injection. The injection should be made into the distal end of the vein thus ensuring that a proximal portion of vein remains for further injections should the initial injection fail distally. Any flow of blood from the vein subsequent to the injection can be stemmed by brief pressure with a piece of cotton wool. The rabbit usually has to be restrained during the injection procedure (Fig. 8).

In the guinea pig, the best vessel in the ear for injection lies medially and is small. It is most conveniently dilated with xylene or toluene. The injection procedure is somewhat difficult in comparison with that for the rabbit but it is worth trying as an initial procedure, especially as anaesthesia is not required.

C. Penile vein. (M, Rt, Gp). Under anaesthesia, the glans penis is exposed by retraction of the prepuce (fore-skin) and is held with the thumb and index finger. The large ventral penile vein can be entered with a 26 gauge (or smaller) needle, depending on the size of the vein.

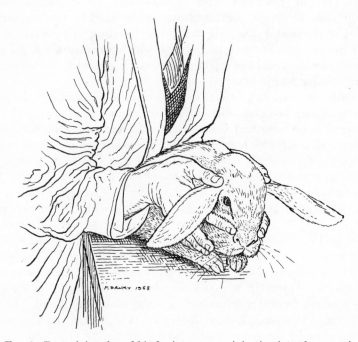

FIG. 8. Restraining the rabbit for intravenous injection into the ear vein.

D. Femoral vein. (M, Rt, Gp, Rb). The animal under anaesthesia is placed on its back and one hind limb is maintained in a stretched position (a rubber band or a piece of string can be used). The hair is clipped short, the leg is swabbed with a 1% (w/v) Cetrimide solution (or a similar detergent/antiseptic solution) and a small mid-line incision made in the region of the thigh. The femoral blood vessels should be located immediately beneath the incision, being situated towards the inside of the leg. The blood vessels are covered by several layers of thin connective tissue and careful removal of these facilitates entry into the vein. The vein is dilated by applying finger pressure in the groin. After entering the vein, the pressure in the groin is released and the injection is made using a 26 gauge needle. Before closing the incision with skin clips or by suturing, a cotton-wool haemostat is briefly applied to the injection site.

E. *Jugular vein.* (M, Rt, Gp, Rb). This procedure is carried out under anaesthesia. A small longitudinal or transverse skin incision is made in the base of the neck, just medial to the proximal end of the fore-limb. The large external jugular vein may be covered with fat which is carefully removed. Farris and Griffith (1966) suggest entering the vein with the needle pointing towards the head by passing through the edge of the pectoral muscle to prevent bleeding. The needle is then turned in the vein so that it points towards the heart. A 26 gauge needle can be used and the injection is made very slowly because of the close proximity to the heart.

F. *Tail vein.* (M, Rt). Injection into the four equatorially placed tail veins has always been considered difficult, principally because they are small, cannot be seen and because the skin is tough and not easy to penetrate. Ronai (1966) has described a procedure which is said to make the injection particularly easy and to result in a very high success rate. The animal is restrained either in a towel (see Faris and Griffith, 1966) or in a tube or box which restricts movement. The tail is then dipped into a 10% (w/v) aqueous solution of sodium sulphide which has a pH of about 11·7. After immersion, (mouse 3 min, rat 5–10 min) the tail is wiped against the direction of hair growth with gauze and then washed with water. This procedure removes the hairs and the superficial layers of the epidermis and also softens the skin and greatly dilates the blood vessels. Injection is made conveniently with a 26 or 30 gauge needle into the lateral veins which are the most prominent. The tail veins are said to remain dilated for up to 24 h so that the preparation of the tail can be carried out sometime prior to the injection.

G. *Injection of neonatal rats and mice.* A technique has been described for the intravenous injection of newborn mice and rats by Andersen et al. (1959). It involves the use of a syringe, held in a simple foot-operated, motor-driven, continuous flow injection apparatus, which is attached to a 30 gauge needle by polyethylene tubing. The injection is made through the outer lateral edge of the unopened eyelid into the subcutaneous infraorbital vein, which runs from the eye to below the ear and then down the neck (Fig. 9). Sighting of the veins is facilitated by wiping the area with glycerol and gently compressing the thorax which increases the central venous pressure. The failure rate is said to be as low as 5%.

Manual injection is possible but generally results in a high failure rate, principally because of the ease with which the vein can be pierced during discharge from the needle. Anaesthesia is not used in this technique.

5. Intraoral

A. Mouse and rat. A syringe fitted with a blunted hypodermic needle is used for the administration. The needle should be of large bore and of sufficient length to pass completely into the stomach. This has the advantage that accurate intragastric dosage is assured and accidental discharge into the lungs can be avoided. If insertion is intratracheal in error, then about half the needle will be showing above the mouth and a fresh attempt must be made. For adult rats, a needle of length 11 cm and with an external diameter of 1·5 mm (15 gauge) has been found

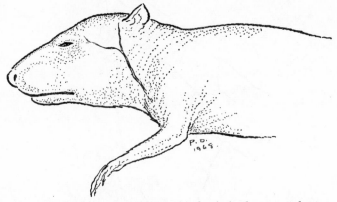

FIG. 9. The subcutaneous infraorbital vein in the neonatal rat.

useful. For mice, a 21 gauge needle will suffice. The use of a No. 3 gum catheter has been suggested for use with rats (see Johnston, 1959) since such a catheter apparently cannot pass into the trachea. A convenient method for holding the rats is shown in Figs 3 and 10. Light but complete anaesthesia, while not necessary, greatly increases the ease with which the needle can be inserted. This is particularly welcome when large numbers of animals have to be dosed.

The mode of introduction of the needle into the stomach is illustrated in Fig. 10. The syringe is held with the thumb and second and third fingers and the needle is passed into the mouth as far to one side as possible. The passage of the needle may be obstructed at two places, viz. at the back of the mouth and at the sphincter to the entrance of the stomach. Manipulation of the syringe to produce a gentle thrusting movement and also perhaps a partial removal of the syringe and its reinsertion slightly more dorsally, especially at the sphincter, will overcome these difficulties. Discharge from the syringe can be carried out fairly rapidly.

B. Guinea pig and rabbit. The animal should be held by an assistant.
A gag, which can be made from a flat piece of hardwood and which
contains a centrally placed hole, is positioned in such a way that the
mouth is forced open and the tongue is pushed forward (Fig. 11). A
polyethylene tube of about 2 to 3 mm in diameter and attached to a
syringe is passed into the stomach via the hole in the gag. The length of
tubing required to reach the stomach should be first noted so that
accidental intratracheal intubation is avoided.

Fɪɢ. 10.   Method of intraoral administration in the rat. The hypodermic needle
should be of just sufficient length to pass into the stomach.

III. METHODS OF OBTAINING BODY FLUIDS
A. Blood
1. Cardiac Puncture

A. Mouse and rat. The animal is anaesthetized and placed on its back
with its body positioned right and left of the investigator. The thumb
and fingers are placed on either side of the thorax which is then slightly

compressed at its base. A hypodermic needle of 21 gauge (rat) or 24 gauge (mouse), attached to a syringe, is inserted under the xiphoid cartilage which has been raised by the index finger and slowly pushed into the pleural cavity while slightly angled to the surface of the body

FIG. 11. Method of restraining the rabbit for intraoral administration.

(Fig. 12). The position of the top of the heart can be gauged when the syringe, which initially should be held lightly, begins to transmit the heart beat. At this juncture the syringe is raised slightly and slowly pushed into the heart. Occasionally the force of the blood will automatically move the syringe plunger backwards but it is more usually necessary to aspirate the blood. The plunger can either be withdrawn by holding it with the thumb and second finger and using the index finger to steady the barrel or by holding the barrel with one hand and pulling the plunger back with the other. If blood cannot be withdrawn, then the heart may either have been completely pierced, when slow withdrawal of the syringe will bring the needle back into the heart cavity, or it may have been by-passed, when a fresh attempt should be

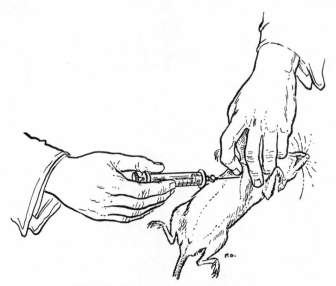

FIG. 12. Cardiac puncture in the rat.

made. Anaesthesia should be continued throughout the procedure. Withdrawal of blood from the heart should be made at the rate of about 10 ml/min or less. The average blood volume of the mouse and rat is 64 ml and 78 ml per kg of body weight, respectively (Altman and Dittmer, 1964). About 50–75% of the blood volume can be withdrawn at any one time while still allowing recovery of the animal. Cardiac puncture can be carried out at weekly intervals if required but a smaller time space may be hazardous to the animal.

B. *Guinea pig and rabbit.* Cardiac puncture is generally performed on the conscious rabbit which is held on its back as in Fig. 13. An assistant holds the front legs forward and steadies the head, while the investigator's forearm is placed along the ventral surface of the animal, thus preventing the animal from righting itself. The thorax is palpated until the point of strongest heart beat is located. This is then marked by the tip of the thumb on one side and by the index finger on the other. A

21 gauge needle attached to a syringe is inserted in between the ribs at a point immediately below the thumb and pushed slowly towards the index finger. The syringe should be aspirated occasionally during this procedure until blood flows into the barrel indicating that the heart has been entered. Great care must be taken that the animal is not allowed to move during the withdrawal of blood otherwise a fatal haemorrhage may occur.

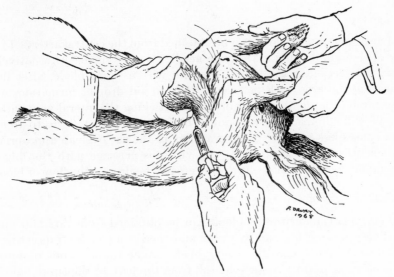

FIG. 13. Method of restraining the rabbit for cardiac puncture.

The same procedure is used for the adult guinea pig but it is usually carried out on the anaesthetized animal. The sub-xiphisternal approach can be used in small animals of this species.

The average blood volume of the rabbit and guinea-pig, respectively, is 56 ml and 75 ml per kg of body weight (Altman and Dittmer, 1964).

2. Jugular Venepuncture

The same procedure is used as is described for injection into the external jugular vein. The needle is passed into the vein pointing towards the head of the animal. Mouzas and Weiss (1960) have described the method in detail for rats and mice and claim that large volumes of blood may be removed in this way, especially from mice, without haemolysis. Kassel and Levitan (1953) perform jugular venepuncture in the conscious mouse using a 26 gauge needle and shaving the neck to sight the jugular veins through the skin.

3. Withdrawal from the Inferior vena cava

The animal is anaesthetized and the abdominal cavity is quickly opened. The inferior *vena cava* is located and entered just below its junction with the renal veins. Blood is withdrawn either by syringe or, in the rabbit, by the use of a vacuum pump attached to a needle by a length of siliconized polyethylene tubing. The procedure results in the animal's death.

4. Ear

Large amounts of blood can be obtained from the central artery of the ear of the rabbit using a 21 gauge needle and a syringe. Alternatively, semi-transection of the lateral ear vein with a sharp blade, after first smearing the area with petroleum jelly to aid droplet formation, will also produce large quantities of blood. Pricking the lateral vein with a needle will allow a small amount of blood to be collected (up to 10 ml) and causes less trauma. Prior warming of the ear over an appropriate heat source (e.g. a lamp bulb) and vigorous massage with the fingers should always be carried out as this greatly facilitates the flow of blood.

5. Tail

Very small quantities of blood can be obtained from rats and mice by first warming the tail in water and then completely transecting a small distal portion with a sharp blade. Anaesthesia is not required. It is often helpful to "milk" the tail from the base to the tip to aid the flow of blood.

6. Decapitation

The animal is stunned by a blow to the back of the head and the neck is entirely or partially transected with scissors or shears. The blood normally spurts out and should therefore be collected in a wide-mouthed container. The blood may be contaminated from contact with the surrounding tissues.

B. Urine

I. Metabolism Cage

This is usually a cage with a wire-mesh floor through which the urine and faeces can pass and which can be suspended over a funnel of suitable size. Urine and faeces are collected by the funnel and separated by means of a suitable apparatus. A metabolism cage arrangement which may be used for rats and guinea pigs is illustrated in Fig. 14. Separation occurs on the basis that the faeces drop through the apparatus into the beaker while the urine runs down the surface of the glass and is channelled out through the side-arm. Contamination of the urine by food and water can occur unless some provision, such as separate compartments, is made to prevent this. Metabolism cages suitable for all purposes can be obtained commercially.

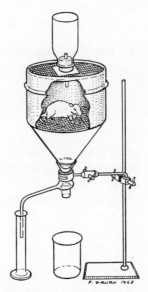

Fig. 14. A metabolism cage for use with rats and guinea pigs. Faeces pass into the beaker whilst the urine is channelled out via the side-arm.

2. Withdrawal from the Bladder

The bladder of the male rabbit, but rarely that of the female, can be catheterized with polyethylene or rubber tubing by passage through the urethra. The catheter should be smeared with glycerol before it is introduced into the penis.

Urine can also be obtained directly from the bladder of any animal by aspiration with a syringe after opening up the abdominal cavity under anaesthesia.

C. Mammary Gland Secretion

Milk can be obtained by simple intermittent suction of the nipples using polyethylene tubing attached to a suitable container which is connected to a vacuum pump. The area around each nipple is first greased with petroleum jelly and the mammary gland is massaged towards the nipple undergoing suction. The greatest yields of milk are obtained when the young are first prevented from suckling the mother for about 24 h. About 30 min before suction is applied, the animal is given an intraperitoneal injection of oxytocin (1–10 i.u. per animal, depending on size) to facilitate the ejection of the milk from the glands.

Temple and Kon (1937) have described a simple electrically operated apparatus for milking rats which yields up to 8 ml/animal. Kahler (1942) has described an apparatus for use with mice. Guinea pigs are best milked by hand but this is rarely successful with rabbits.

D. Lymph, Bile, Pancreatic Secretions

These are dealt with in Section IV.

IV. SURGICAL TECHNIQUES IN THE RAT

A. Introduction

The surgically prepared animal is being used increasingly in biochemical research, especially that aspect dealing with the influence of humoral substances on the biosynthesis of proteins (Tata, 1967). The present section deals with a number of surgical techniques applied specifically to the rat which continues to be the most popular experimental animal. Some selection inevitably has to be made but it is hoped that the methods described cover the major requirements of research work today.

There can be no doubt that animal surgery demands considerable skill if homeostatic interference is to be kept to the minimum. Expertise in this field brings its own rewards, such as the longer survival of animals and the production of an animal which is healthier and which responds better and more accurately to subsequent treatment. A good knowledge of anatomical relationships is essential as this avoids the accidental manipulation of extraneous organs and tissues which could result in unwanted complications such as haemorrhage or nerve damage. The precise surgical technique adopted is often a matter of personal choice and the skill of the operator. In this respect, therefore, modifications of the techniques described in this section are to be expected and indeed welcomed as an indication of the degree to which the operator has become involved.

The rat provides excellent surgical material for three main reasons: it is of a convenient size, it requires little in the way of complicated apparatus and it is highly resistant to infection. Because of this latter attribute, it is unnecessary to preserve asepsis during an operation, but it must be emphasized at the same time that clean environmental conditions, clean instruments and clean hands should be employed as a matter of good practice. The preparation of the animal prior to surgery takes the form of a standard procedure irrespective of the type of operation to be performed. The animal is anaesthetized (see below) and placed on an operating board of soft wood or of cork. The area of the skin into which the incision is to be made is shaved or the hair is clipped short (many workers omit this procedure) and is then swabbed with either industrial methylated spirits, 74 over-proof, or with a detergent/ antiseptic solution such as 1% Cetrimide. This process prevents the animal's hair from interfering with the operation and also affords a certain degree of asepsis. Anaesthesia is continued throughout surgery and its maintenance is best relegated to an assistant. The instruments needed for surgery are simple in nature, the basic requirements consisting of two medium-sized pairs of scissors, one pair blunt-ended and the other pointed; three pairs of curved forceps, one pair being of a fine calibre; surgical needles; cotton or silk thread and Michel skin wound clips. Skin and muscle incisions are always made with scissors and the scalpel is rarely used in rat surgery. Closing incisions in muscle and other tissues has become a precise art in veterinary and human surgery and some knowledge in this field is generally helpful for the rat. A useful insight into this aspect of surgery is given by Leonard (1968). In the majority of cases, special pre-operative and post-operative care is rarely necessary but should these be called for they are indicated below under the individual techniques.

B. General Anaesthesia

I. Volatile Anaesthetics

A. Ether. A number of volatile anaesthetics suitable for use with small laboratory animals are available. By far the most widely used of these is peroxide-free (anaesthetic) ether. It is comparatively inexpensive and is rapidly effective, the length of sleep being controllable and possible of being maintained for a period of 2 h or more. The animal apparently recovers from its effects very quickly, although some of these effects may be deleterious in biochemical work, e.g. ether is known to affect the ultra-structure of the liver.

To induce anaesthesia, the animal is placed in a confined space, e.g. a covered glass jar or an air-tight box. Into this is placed a wad of cotton wool soaked in ether followed by a wad of dry cotton wool to prevent the animal from suffering unnecessary irritation due to direct contact with the ether. Anaesthesia is rapid. The operator should become familiar with the signs showing that the required level of anaesthesia has been reached. In general, the animal is completely anaesthetized when the muscles of the head and body have become relaxed. (To determine this state it is sufficient to tilt the container until the animal falls over and to watch for any effort it makes to raise its head or to move.) Movement of the vibrissae should have stopped and the animal's respiration should be regular. At this stage the animal may be removed from the container and placed on the cork operating board when two further tests can be carried out. These are the testing of the corneal reflex, when the eye blinks as the cornea is touched, and the pedal reflex, when the foot jerks on pinching the toes. If both of these reflexes are absent, adequate anaesthesia has been induced.

The maintenance of anaesthesia during surgery is carried out by intermittent dropwise application of ether either on to a strip of gauze or on to a gauze-covered wire cone which is placed over the nose and mouth. The ether should be placed around and not onto the nose, which will respond unfavourably with mucous secretion if wetted. In some cases (e.g. during hypophysectomy—see later) it is advantageous to continue anaesthesia using tracheal intubation. This is simply performed by exposing the trachea (see under thyroidectomy) and inserting a large-bore hypodermic needle (e.g. 18 gauge) or a needle-tipped polyethylene tube. The end of the needle or the tube is then covered with a piece of gauze to which ether is applied. A more refined method of intratracheal ether anaesthesia has been described by Farris and Griffiths (1966).

Over-anaesthetizing must be avoided at all times. If respiration stops as a result of over-anaesthetizing, resuscitation must commence immediately using either one or a combination of the following procedures. The tongue should be pulled out and extended. At the same time, the mouth must be kept open by expanding forceps between the lower and upper jaws. The thorax should then be compressed intermittently while blowing vigorously into the animal's mouth. This method alone is often successful. An alternate procedure is to suspend the animal by its forelegs and swing it in such a way that the body "jacknifes", thereby causing efficient compression of the thorax. The most difficult cases will often respond after gentle, intermittent inflation of the lungs by blowing down a polyethylene tube placed in the trachea either by way of the glottis or by tracheotomy. Attempts at revival are worth continuing for as long as a strong heart-beat can be felt.

The major disadvantage of ether as an anaesthetic is the excessive salivation and bronchial secretion that it causes. To avoid this, particularly in long operations, atropine sulphate can be administered subcutaneously 15–30 min prior to the operation at a dose of 16 mg/ 100 g of body weight. Apart from inhibiting nasopharyngeal mucous secretion, atropine is also a circulatory and respiratory stimulant and causes dilation of the bronchioles, all of which attributes are helpful during anaesthesia.

B. *Trichloroethylene*. This anaesthetic has a low volatility which should be borne in mind when it is used dropwise, otherwise over-anaesthetizing is apt to result. It induces sleep rapidly and is said to reduce haemorrhage and salivation. It is used in a similar manner to ether.

2. Barbiturate Anaesthetics

A. *Pentobarbitone sodium*. Good anaesthesia lasting for 1 h or more can be induced by a single intraperitoneal, intramuscular or intravenous injection of pentobarbitone sodium. The intraperitoneal route is the one most frequently used. A dose of 30–50 mg/kg of body weight, depending on the depth of anaesthesia required, produces sleep in up to 20 min. Commercially available, ready-prepared veterinary Nembutal can be used conveniently for the intraperitoneal or intramuscular routes. For intravenous administration, the same dose should be prepared by dissolving the powder in either distilled water or physiological saline. The injection can be given either into the tail vein of the conscious animal, or into one of the other accessible veins of an animal lightly anaesthetized with ether.

B. *Sodium amytal*. This anaesthetic is relatively unstable in solution and should be made up freshly prior to administration. The dose for the rat is 60 mg/kg of body weight injected intraperitoneally.

It is worth noting that some discretion must be used when administering non-volatile anaesthetics. The anaesthetic dose is often close to the lethal dose and it is not unusual for the occasional death to occur during anaesthesia. It must also be noted that resuscitative procedures are usually ineffectual in animals given barbiturates. Most workers compromise, therefore, and administer one-half the maximum recommended dose, which is sufficient in itself to produce narcosis, and deepen the sleep by the very sparing use of a volatile anaesthetic such as ether. Whenever intravenous anaesthesia is used, the animal must be watched carefully for signs of sleep during the injection. It should be tested frequently for the absence of the corneal and pedal reflexes. The injection, which must be made slowly, is stopped irrespective of the amount of anaesthetic given as soon as the reflexes cannot be obtained. A lightly etherized animal, which may not show the normal reflexes, should be given one-quarter of the normal intravenous dose of barbiturate in order to continue the anaesthesia adequately.

C. Partial Hepatectomy

The anatomical relationship of the blood vessels supplying the liver
is illustrated in Fig. 15. In the usual form of this operation the bifur-
cated median lobe and the left lobe of the liver are removed together
since they have a common blood supply. These two lobes constitute
about 70% of the mass of the liver.

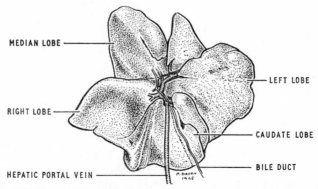

Fig. 15. The rat liver with the hepatic portal vein and the bile duct.

For the operation, the animal is laid on its back and a ventral mid-
line incision about 2 cm long is made posteriorly from the edge of the
xiphoid cartilage, first through the skin and then through the abdominal
muscles. At this juncture, the animal is repositioned left and right of
the operator and a 3 cm diameter cotton wool–gauze bolster is placed
under the thorax causing the liver to fall forwards. The transparent
suspensory ligament attaching the convex face of the liver to the diaphragm is severed down to the inferior *vena cava* with blunt curved
scissors. There is no necessity to remove the xiphoid cartilage if care
has been taken to start the incision just below it so that it is not
exposed. A thick piece of gauze is placed along the nearside edge of the
incision and the median and left lobes of the liver are moved out of the
abdominal cavity onto it. Mobilization of these lobes is neatly effected
by pushing the gut just posterior to the liver forwards and upwards
in a concave semi-circle with simultaneous light compression of the
abdominal contents.

The liver lobes to be excised can then be easily but gently pulled out
and laid convex side down onto the gauze. The slight tension produced
in these lobes after mobilization will reveal two other suspensory
ligaments, one attaching the concave face of the left lobe to the blood
vessels and to the stomach and one attaching the antero-lateral edge of

the same lobe to the dorsal peritoneum. Both these ligaments must be cut. The two liver lobes can now be raised vertically and a ligature tied around them and the blood vessels at their base. The lobes are again laid on the gauze and several cuts are made in them peripherally to allow bleeding. The gauze is folded over the lobes which are then picked up, placed under slight tension and severed with fine-pointed scissors as near to the ligature as possible. The muscle incision is sutured and the skin incision is closed with Michel wound clips. Mortality from the operation is extremely low. Regeneration of the liver normally occurs within 14 days. The main features of the technique are illustrated in Fig. 16.

D. Ovariectomy

The animal is laid on its ventral surface with its tail towards the operator. A mid-line dorsal skin incision is made approximately half-way between the dorsal insertion of the last pair of (floating) ribs and the base of the tail. The landmark to look for here is the peak of the natural curvature of the spinal cord. The peak of this "hump", easily seen with the animal in this position, occurs at the point at which the terminal pair of ribs join the last thoracic vertebra. Pointed scissors are alternately inserted subcutaneously through the single incision, down each side of the animal and opened so as to spread and tear the connective tissue allowing direct access to the muscle. The skin incision can then be retracted laterally first to one side to remove one ovary and then to the other side to remove the second ovary. Entrance to the peritoneal cavity on either side of the animal, in the region where the ovaries lie, is gained by making a small cut with fine scissors into the muscle about midway between the dorsal and ventral surfaces. This incision should be several millimetres below the spinal muscles. It is good technique to make the muscle incision just wide enough to allow the passage of the ovary. In practice, accurate control of the width of the incision is obtained by cutting only partially into the muscle with scissors (i.e. scoring it) and then entering the peritoneal cavity by forcing the points of a pair of curved forceps through the remaining layers of muscle. The width of the hole so produced can be regulated by expanding the forceps. If the muscle incision has been correctly placed, the ovary surrounded by a variable amount of fat will be found underneath or within easy reach. In old rats the ovary is often totally obscured by fat but this usually acts as a landmark and should present no problems since it is freely movable. This should not be confused with the fat which will be found under and attached to the spinal muscles and which is not freely movable.

To remove the ovary, it is pulled through the muscle incision by grasping the periovarial fat. The ovary itself must not be touched otherwise small pieces may become detached and re-enter the abdominal cavity where they may implant and carry on normal function. With fine scissors, the junction between the oviduct and the uterine horn,

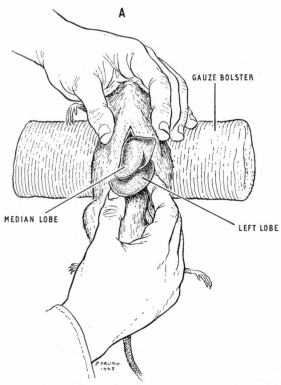

FIG. 16. Stages in partial hepatectomy in the rat. A. Mobilization of the median and left lobes of the liver out of the abdominal cavity. B. The hepatic suspensory ligaments of the left lobe which are sectioned during the operation. C. Placing a ligature around the blood vessels. D. Cutting the ligated liver lobes on gauze to allow bleeding prior to their removal.

together with all accompanying blood vessels, is severed and the horn is returned into the abdominal cavity. In the rat, irrespective of age, there is rarely any necessity to observe haemostasis during the operation. Bleeding is usually small and soon stops of its own accord. The muscle incision requires no suturing unless it has been made inordinately large when a single suture will suffice. The skin incision is closed with wound clips.

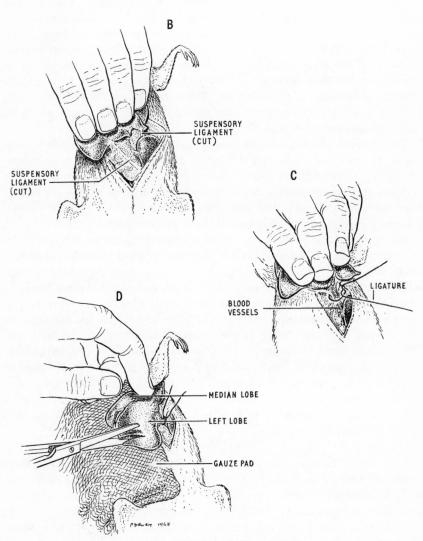

FIG. 16 B, C, D.

E. Adrenalectomy

The procedure for gaining entrance to the peritoneal cavity is similar to that described for ovariectomy (see above) with two important modifications. Firstly the skin incision should be more anterior and slightly below the peak of the "hump" of the back (i.e. about one-quarter the distance between the last pair of ribs and the base of the tail). Secondly, the muscle incision on the right side of the animal, for removal of the right adrenal, is made immediately beneath the spinal muscles rather than several millimetres below them and is placed more anteriorly to that on the left side. These modifications correlate with the anatomical positioning of the adrenals which are found medially and near the anterior poles of the kidneys.

The correct placing of the muscle incision on the left side should allow the anterior portion of the spleen to be seen directly underneath or slightly to the left. A pair of curved forceps is then inserted into the incision so that the spleen lies laterally to the forceps and is prevented from moving medially and obscuring the adrenal. A second pair of forceps is inserted fairly deeply and the incision together with both pairs of forceps is retracted medially. This manœuvre is designed to push the greater part of any loose fat out of the area around the adrenal and to bring the incision to lie over the gland. Some deep searching for the adrenal gland may be necessary if fat is present, since the fat tends to cause the gland to lie dorsally and may totally obscure it. Once found, the gland is brought out through the muscle incision by holding the periadrenal fat, care being taken to avoid touching the gland itself. Removal is effected by clamping the tissues and blood vessels at the base of the adrenals with both pairs of forceps as in Fig. 17. No further attempt at haemostasis is necessary; any bleeding is usually small and of no consequence. The forceps are then drawn sharply away from each other and any fat or other tissues, minus the adrenal, is returned into the abdominal cavity.

On the right side, the adrenal gland and the anterior pole of the kidney are overlain by the peripheral part of the liver. This, therefore, is pushed away anteriorly by the convex surface of one pair of forceps and the adrenal gland is withdrawn by a second pair. The removal procedure is identical to that for the left side.

In immature or young adult rats, little fat will be present to obscure the adrenal glands on either side. Consequently if such animals are to be consistently used, the muscle incision for removal of the left adrenal can be made at the border of the spinal muscle, as is done on the opposite side. The gland should here lie superficially and directly under-neath. The operator will have to decide which method suits him best.

No suturing of the muscle incision is required if this has been kept small and the skin is closed with wound clips. The experienced operator should find no difficulty, using young rats, in removing up to 100 pairs of adrenal glands per hour!

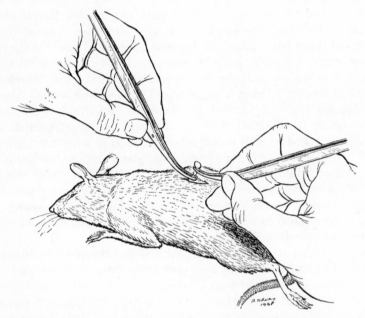

Fig. 17. Adrenalectomy in the rat.

F. Orchidectomy

The animal is laid on its back. A median incision about 5 mm in length is made through the skin at the tip of the scrotum. The subcutaneous connective tissue immediately encountered is cleared and it will then be seen that each testis lies in a separate muscular sack. An incision is made into each muscular sack at its base and the *cauda epididymis* is pulled out accompanied by the testis. A ligature is placed around the *vas deferens* and the spermatic blood vessels at the anterior pole of the testis and the glands are removed by severing below and close to the ligature. The skin incision is closed by one or two Michel clips.

G. Thyroidectomy

A mid-line ventral incision about 2 cm long is made in the neck between the sternum and the widest part of the lower jaw. The subcutaneous fat and connective tissue are cleared and the large, pink-coloured salivary glands are retracted laterally by grasping the connective tissue between them with two pairs of forceps and then pulling sideways away from each other. The sternohyoid muscle, which covers the trachea, is split along the middle by running the points of a pair of fine curved forceps up and down until the median fascial plane is revealed. The two halves of the muscle are then separated using retractors (bent paper clips can be used effectively here). Note that a narrow strip of muscle, joined posteriorly to the sternohyoid muscle, runs in close association over the thyroid gland and must, therefore, be separated from it. Using a single-edged irredectomy scalpel, the isthmus joining the two thyroid glands across the trachea is carefully transected in the middle. With fine forceps, the "free" portion of the isthmus is picked up and pulled laterally, placing tension on the gland itself. The connective tissue between the gland and the trachea is carefully cleared while the gland is under lateral tension, using the blunt edge of the scalpel as much as possible. The operator must proceed with extreme caution as the dorsal edge of the gland is approached, since along it runs the recurrent laryngeal nerve which must be left intact, otherwise respiration is fatally impaired. The nerve will be seen as a fine white strand attached loosely to the edge of the gland by connective tissue. The gland is carefully dissected away from the nerve and finally severed from any attachments anteriorly and posteriorly including the nerves to the gland itself. During the dissection, bleeding will inevitably ensue as the blood vessels to the thyroid are cut. Brief pressure with cotton-wool pellets should stop the bleeding and mop up any blood which is obscuring vision. Before closing the skin incision with clips, all retractors and pellets are removed and the muscles are allowed to return to their original positions. No suturing of the two halves of the sternohyoid muscle is necessary. The main steps of the technique are illustrated in Fig. 18.

The parathyroid glands which are to be found embedded in the thyroid at its anterior pole are unavoidably removed by the above technique. The tetany which would ensue because of calcium depletion is counteracted by adding 2% (w/v) of calcium lactate to the drinking water.

H. Hypophysectomy

This is a difficult technique and extensive experience on dead animals should be gained first, using the parapharyngeal approach to the pituitary gland. For the operation, the animal is laid on its back and its neck is stretched out by string passed around the incisors. The trachea is exposed (see Section G) and a large-bore hypodermic needle is inserted into it to facilitate respiration throughout the rest of the operation. Anaesthesia can be maintained if necessary by placing gauze over the end of the needle and applying ether to it. The points of a pair of curved forceps are forced through the muscle at the lower edge of the sternohyoid (to the right of the operator) where it adjoins the digastric muscle. By opening the forceps in all directions the area can be

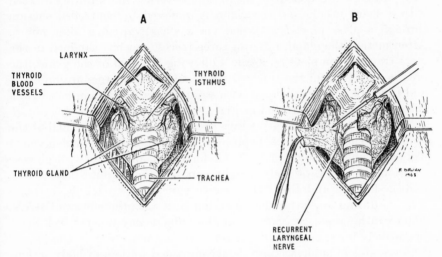

Fig. 18. Thyroidectomy in the rat. A. Exposure of the thyroid gland. B. Dissection of the gland after severing the thyroid isthmus.

widened and cleared. Care must be taken not to sever any nerves or large blood vessels. The sternohyoid muscle, the trachea and the oesophagus are retracted laterally. It is very important during the latter part of the operation to keep the working area as open as possible by the liberal use of retractors. The forceps are now inserted deeply, separating any muscles that lie in the way, until the white under-surface of the cranium (occipital bone) is reached. The posterior part of this area of the cranium is bounded by a small depression, produced by the occipital foramen, which is several millimetres posterior to the floor of the fossa where the pituitary gland lies. At this stage of the operation a dissecting microscope will be found useful. The surface of the skull is scraped clean of muscle and other tissue with the points of forceps, working in an anterior direction until a slight upward slope in the bone is encountered. At the top of this, a transverse suture, "blue" in colour, will be seen. The pharynx lies very close at this point and care must be taken not to pierce its wall with the forceps. During the blunt dissection, blood should be mopped up with cotton-wool pellets. Entrance into the skull is gained by using a foot-operated dental drill to which has been attached a round bone burr of suitable size (size 9 is adequate for rats of body weight 150–200 g). A hole is drilled very carefully so that at least two-thirds of it lies anterior to the suture line. The hole must be accurately centred otherwise fatal haemorrhage will occur. Once through the bone, the membrane covering the pituitary is torn with a sharp needle and the gland is removed by controlled suction through a glass pipette attached via a glass trap to a filter pump. After suction the gland, often in three pieces, can be examined in the glass trap for complete removal. Following removal of the gland, the area is wiped clean, the tracheal tube and the retractors are removed and the skin incision is closed with wound clips. The main features of the operation are illustrated in Fig. 19.

Attention to post-operative care is essential for good survival of the animals. They should be kept in a warm room (28°) and given a 5% (w/v) glucose solution to drink (or 0·5 ml of a 5% (w/v) glucose solution per 100 g of body weight can be injected intraperitoneally daily). Falconi and Rossi (1964) give cortisone (1 mg/100 g of body weight) during and after the operation to reduce the effects of stress. Food can be given ad libitum but the animal may not eat well postoperatively for one or two days.

Successful hypophysectomy is accompanied by loss of body weight and atrophic changes in the thyroid gland, the gonads and in the adrenal cortex. Completeness of hypophysectomy should always be confirmed either macroscopically at autopsy or, more accurately, by the microscopical examination of serial 10 μ sections of the *sella turcica* which houses the gland in the skull.

In recent years there has been a revival of the intra-aural approach to pituitary ablation. Removal of pituitary glands in young animals at the rate of up to 90 per h has been claimed. For details of the procedure, the reader is referred to the work of Falconi and Rossi (1964) and of Sato and Yoneda (1966).

I. Lymph Duct Cannulation

Most descriptions of this difficult procedure have referred to the cannulation of two lymph ducts; the intestinal duct which is closely associated with the superior mesenteric artery and which opens into the *cisterna chyli* and the thoracic duct which drains the cisterna in the abdomen and travels through the thorax into the cervical region. The cannulation of only that part of the thoracic duct which is found in the

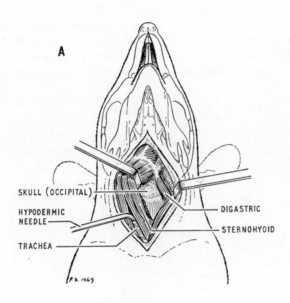

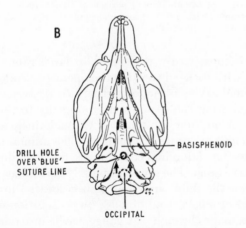

FIG. 19. Hypophysectomy (parapharyngeal route) in the rat. A. First sighting of the skull directly beneath the incision between the sternohyoid and digastric muscles. B. Ventral aspect of the skull. The region of the skull first sighted is indicated by a broken circle with arrows showing the direction of the dissection to reveal the "blue" suture line. The drill hole to expose the pituitary gland is made just anterior to this suture line.

abdominal cavity will be described (see Fig. 20). A description of other
procedures has been given by Lambert (1965).

The abdominal portion of the thoracic duct lies close to and slightly
beneath the dorsal aorta and arises about 2 cm below the diaphragm
at about the level of the left adrenal vein. It passes underneath the
aorta as the *cisterna chyli* and gives rise to the intestinal lymph duct.

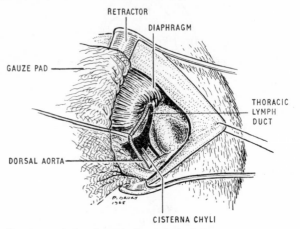

Fig. 20. Exposure of the abdominal portion of the thoracic lymph duct. The
dorsal aorta is retracted with cotton thread. The gut, the right kidney and the right
adrenal gland have been reflected to the left.

The operation is most easily performed in large rats. The animal is
laid on its back with its head towards the operator. An oblique incision
into the abdominal cavity is made starting just posterior to the xiphoid
cartilage and following the costal margin dorsally to the border of the
spinal muscles. A mid-line incision can be made instead but this may
result in insufficient exposure of the duct. The animal is turned on its
right side so that the abdominal contents are reflected downwards away
from the operative area. The group of pink lymph nodes situated at the
junction between the ilium and the colon is located in the mesentery
and an injection of 0·05–0·1 ml of a 1% (w/v) solution of Trypan Blue
or Evans Blue is made through a 30 gauge needle into one or more of the
nodes found in the middle of the group. Reinhardt (1945) suggests, as
an alternative, the intraperitoneal injection of 0·5 ml of the dye 30 min
prior to the operation. The injection of the dye colours the lymph blue
within seconds and allows easy recognition of the abdominal lymph
ducts. It remains in the lymph circulation for a short time only.
The gut and the liver, freed of ligamentous attachments, are packed
well into the right flank using gauze soaked in warm physiological

saline. If possible the left kidney and adrenal gland should also be reflected to the right as this facilitates the subsequent dissection. The liberal use of retractors to keep the operative field widely open is of paramount importance during the operation. A small opening in the dorsal peritoneum should be made slightly anterior to that part of the duct which crosses under the aorta. By blunt dissection with the points of a pair of forceps, the duct is separated from the aorta, care being taken not to sever the intercostal arteries which intermittently cross over the lymph duct. The duct itself should be touched as little as possible as it is delicate and tears easily. The aorta should be retracted laterally using a loop of thread as this allows better visibility of the duct. A ligature, using fine thread, is placed around the duct about 5 mm from the diaphragm. This can be done by using fine curved forceps to pull the thread under the duct. The ligature stops the anterior flow of lymph and causes the duct to double in size within a few minutes. A second ligature, untied, is placed just above the *cisterna chyli*, about 10 mm below the first. A polyethylene cannula of 0·6–1·5 mm external diameter is then passed into the duct. In the method used by Bollman *et al.* (1948), the cannula, filled with dilute heparin, is placed via a trocar needle through the abdominal muscles near the xiphoid cartilage. The curvature of the diaphragm is then followed until the cannula comes to lie in the same plane as the duct. Using the cutting edge of a 27 gauge needle, a small longitudinal slit is made in the duct between the two ligatures and the bevelled end of the cannula is inserted for a short distance. This procedure is critical since the duct collapses when the slit is made and can very easily tear irreparably. When the tube has been inserted, it is secured by tying the second ligature around it. It is further anchored by tying around it the free ends of the first ligature. The cannula is sutured to the abdominal muscle to prevent external movement being transmitted internally and the muscle and skin incisions are closed in the usual way.

Lymph should pass along the tube as soon as cannulation is effected. Its rate of flow has been reported to be about 0·13–0·7 ml/h. (Lambert, 1965). Lymph clots which may form in the cannula can be broken up with a polyethylene or fine wire stylet. During lymph collection, the animal should be restrained either by a long-acting anaesthetic or by use of the simply made close-fitting restraining apparatus, first described by Bollman (1948).

J. Bile Duct Cannulation

Pure bile or pancreatic secretions can be obtained by cannulation of strategic parts of the bile duct. The region of the duct situated up to

10

about 10 mm from the hilum of the liver where the duct bifurcates is used for collecting bile and cannulation of the region distal to this will collect pancreatic secretions. The anatomical relationships of the bile duct are illustrated in Fig. 15.

Exposure of the bile duct is made through a long post-xiphisternal, mid-line ventral incision into the abdominal cavity. The median and left lobes of the liver are retracted either against the diaphragm or onto the thoracic wall. The duodenal loop, immediately following the stomach, is laid out on wet gauze to the right of the operator. The whole course of the bile duct and its pancreatic connections can now be seen from the hilum of the liver to its point of entry into the lower part of the duodenum. Lambert (1965) suggests placing a 2 cm bolster under the thorax to facilitate exposure of the duct. A ligature is placed around the duct about 5–10 mm from its bifurcation into the liver and above the highest pancreatic duct. The bile duct becomes turgid but shows little enlargement in size. A pair of straight forceps is opened under the duct anterior to the ligature so that the duct is stretched over the points. The duct is semi-transected with fine scissors and at this point it is advisable to use a dissecting microscope. A polyethylene tube of about 0·6 mm external diameter, with a bevelled end, is then pushed into the duct for several millimetres in the direction of the liver. The cannula is secured with a ligature and passed out through a trocar opening in the lumbar region where it is further secured to the skin with a suture.

Harrington et al. (1936) suggest testing for the success of the operation by injecting intravenously a 1% (w/v) solution of Rose Bengal in physiological saline. The normal route for excretion of this dye is through the bile and none should appear in the urine if the bile flow is not obstructed. The flow rate of bile in the rat is approximately 10 ml/24 h (Lambert, 1965).

Collection of pancreatic secretion depends on the cannulation of the bile duct after placing of two ligatures; one near the liver to prevent the flow of bile and one near its insertion into the duodenum to prevent escape of the secretions into the gut. A cannula placed between the two ligatures will thus collect pure pancreatic secretions. Both bile and pancreatic secretions can be collected simultaneously, but separately, by strategically placed cannulas and ligatures. Collection of pancreatic secretions with a simple obstruction of the bile flow causes the animal to become jaundiced within 48 h although animals usually survive in spite of this for one to several weeks. During the prolonged collection of bile or pancreatic secretions the Bollman apparatus (see Section I above) will be found useful for restraining the animal.

V. MISCELLANEOUS TECHNIQUES

A. In Vivo *Liver Perfusion in the Rat*

This technique is used principally to clear the liver of residual blood but, of course, can be used for other purposes where a perfusion of short duration is required.

The rat is anaesthetized and placed on its back. The abdominal cavity is quickly opened by making two oblique incisions each starting in the mid-line at the base of the abdomen and travelling to the dorso-lateral edge of the rib-cage. The triangular flap of skin and muscle so formed is reflected over the thorax; the gut is pulled over to the operator's right side and the large portal vein passing into the liver is located. The vein is brought under tension, by straddling it distally with the index and second fingers of one hand. This prevents the vein from moving forwards when the needle is inserted. Entry is made about 15 mm from the hilum of the liver using a 21 gauge needle attached to a 50 ml syringe filled with physiological saline. After entering the vein, the inferior *vena cava* immediately anterior to the liver is cut with scissors and the perfusion is started. As the blood is displaced from the liver, the latter will become progressively pale in colour. To facilitate removal of blood from the peripheral parts of the liver lobes it is often necessary to massage them gently between the fingers. Subsequent to the perfusion, the liver is removed by gripping the diaphragm with forceps and cutting round it with scissors. After the remnants of the diaphragm have been removed, a liver is obtained which has escaped injury due to direct handling.

B. *The Production of Antibodies*

The rabbit is the animal most commonly used for the production of antibodies to protein or other antigens. There is, however, no reason, other than one of convenience, why other animals such as sheep, goats, horses and turkeys should not be used if it is found more convenient to do so. In many cases it may in fact be advantageous to use one particular species rather than another, since the more dissimilar the protein antigen is to the proteins of the animal to be immunized, the more likely it is for a strongly reacting antibody to be produced.

Careful preparation of the antigen for injection is important if a good antibody titre is to be achieved. The use of adjuvants has found widespread acceptance in immunology because of the long-lasting and high antibody response that can be obtained. The hypersensitive state produced in the animal by the adjuvant is enhanced by the incorporation of various species of mycobacteria and modern adjuvants usually contain one or more of these.

The method of preparation of small quantities of adjuvant–antigen emulsions is as follows. Bayol F, Falba and Arlacel A, all commercially available mineral oils, are warmed until they become free-running and are then mixed in the respective volume ratios of 5 : 4 : 1. Heat-killed *Mycobacterium tuberculosis* or *Mycobacterium butyricium* (1 mg/ml) is then added. The resulting mixture is referred to as complete Freund's adjuvant. The antigen, which is dissolved in distilled water or in physiological saline, is added to the complete Freund's adjuvant in the ratio of 1 volume of antigen solution to 3 volumes of adjuvant. The procedure for producing the final water-in-oil emulsion was first described by Freund and Thompson (1945) and involves recycling the materials through a syringe. The antigen solution, one-fifth of the total volume at a time, is withdrawn into a syringe fitted with a 26 gauge needle and is then quickly expelled under the surface of the oil. When all the solution has been added in this way, the "crude" water-in-oil emulsion so produced is taken up into the syringe via a 21 gauge needle and the whole amount is forcibly ejected through a 26 gauge needle. The white emulsion is recycled in this manner a number of times until it is so stable that one drop placed on the surface of some water in a beaker remains as a discrete drop without dispersion. The final emulsion is extremely stiff and it is advisable to use a Luer–Lok syringe throughout the entire mixing procedure and also for the injection into the animal.

Many authors use a modification of the above procedure whereby only Bayol F and Arlacel A are mixed together in a ratio of 8·5 : 1·5 with the addition of *Mycobacterium*. The "pros and cons" for the use of these two regimens have not caused either one to be considered more favourably at the present time but one advantage of the 8·5 : 1·5 procedure lies in the subsequent emulsification of equal volumes of adjuvant and antigen thus allowing more antigen to be incorporated into the emulsion. This is a useful attribute when proteins of weak antigenicity are being employed. Adjuvants can often be used for up to a year if stored at $-20°$ but should be injected within a few weeks of preparation if stored at room temperature.

The amount of protein to be used for immunization will depend on its antigenicity and its toxicity in the animal. Unfortunately, prior knowledge of these is often lacking and trial and error has to be employed. Amounts of protein in the region of 10 mg are often initially used if nothing is known about its effect in the animal. It is well to note that up to 100 mg of protein can be incorporated into the oil base while still giving a stable emulsion.

The administration of the adjuvant to the animal is made as a single injection, possibly distributed over several sites, either subcutaneously or intramuscularly or, better, by employing both routes simultaneously. The antibody response to this injection reaches its peak usually between 4 to 8 weeks later and thereafter remains at a "plateau" level for up to about 1 year (Freund, 1947). A booster dose of antigen given after 3 weeks usually increases the antibody response. Such boosting can be profitably made with the antigen adsorbed onto alum precipitate to allow a slightly prolonged effect of the antigen. The precipitate is prepared as follows. The protein, in the same concentration or one greater than that used for the adjuvant, is dissolved in 2·5 ml of distilled water. This is added to 1·4 ml of a molar solution of $NaHCO_3$. Two and one-half ml of a 10% (w/v) solution of aluminium potassium sulphate is then added slowly and dropwise, with shaking. The resulting precipitate is centrifuged and washed with 10 ml of 0·067M phosphate buffer of pH 7·3 and finally suspended, after brief homogenization in an all-glass homogenizer, in the required volume of the same buffer. This is administered intravenously to the animal on alternate days for about 3 weeks and is divided into several progressively increasing doses. One schedule that has been used is to give two doses of 0·1 ml, two of 0·2 ml and five doses of 0·4 ml. The serum antibody titre or, alternatively, the ability of the antiserum to form a precipitate with the antigen (see Boyd, 1966) should be determined about 3 to 8 weeks after the adjuvant injection and again at 1 week after the series of booster injections. Providing survival quantities of blood are taken each time it is bled, the animal can be boosted intermittently over a number of months, even years, to produce further quantities of serum with maximum antibody activity.

Recently a technique for the immunization of animals involving the injection of small quantities of adjuvant into the inguinal or popliteal lymph nodes has been described (Newbould, 1965; Goudie et al., 1966). Microgram quantities of antigen are often sufficient to produce a large antibody response using this method.

C. Transplantation of Tumours

A large number of tumours which vary in consistency from solid tumours, such as subcutaneous fibrosarcomas, to fluid tumours, exemplified by the ascites tumours, have been produced in animals. Simple injection with a hypodermic needle and a syringe will suffice to transplant the ascites tumours into various sites including the blood circulation. The same procedure can be used for those "soft" tumours, such as some subcutaneous liver tumours, where the tumour tissue is first minced with fine scissors or pushed through a tissue press to produce a "mash". Here the syringe and needle should have a large bore to allow the material to pass. A Luer syringe with a short 15 gauge needle will be found useful in these circumstances. The standard procedure for transplanting solid tumours involves the use of a trocar, i.e. a needle of very large bore (about 2 mm external diameter) containing a manually operated stylet. The procedure is as follows. The tumour is dissected out, freed of extraneous tissue and cut in half in a Petri dish. It is quickly rinsed in physiological saline and transferred to a second dish. It will often be found that the tumour contains necrotic areas, usually situated centrally, and white, peripherally placed viable tissue. Only this latter is used for transplantation. This tissue is cut out from the tumour with a scalpel and transferred to a third Petri dish wetted with saline. Small blocks of tissue approximately 1 mm by 5 mm in size are cut although the object of this procedure, that of obtaining pieces small enough to fit into the trocar needle, does not make the actual size of the piece of tumour critical. Three or four pieces of the tumour are forced gently into the bevelled end of the trocar and the stylet is inserted for a short distance.

In any transplanting procedure, it is good practice to observe strict sterility. In the present example of transplanting solid tumours, all instruments, including the trocar, should be boiled or sterilized in some other appropriate way and only sterile physiological saline should be used. When the injection is to be made, the site of entry into the skin should be shaved, or the hair clipped short, and it should be swabbed with an antiseptic solution (e.g. proflavine solution). Pus or extensive necrosis in the transplant may call for more stringent observations of sterility.

Once the trocar has been "loaded", it is inserted subcutaneously and the stylet is pushed in its entire length causing the pieces of tumour to be deposited. It is generally accepted that this deposition, irrespective of the type of tumour tissue used, results in a more successful establishment of the tumour if it is made in the axillary region, that is under or near the forelegs, than if made elsewhere, probably due to the abundant blood supply in this area. Successful establishment of a transplant will also depend on the type of animal into which it is injected, the most favourable environment being the same strain of animal in which the tumour originally arose. In some cases, however, the use of different animal strains and even different species will still produce reliable transplanted material. The time taken for the transplanted tumour to "take" in the host animal will vary with the type of tumour and observation should be continued for several weeks should there be no indication of early growth. An excellent description of the methodological aspects of tumour transplantation has been given by Liebelt and Liebelt (1967).

ACKNOWLEDGEMENTS

The author would like to express his deep appreciation to Mr. Peter Drury for providing the illustrations and to Dr. B. Ketterer for reading and criticizing the manuscript.

REFERENCES

Altman, P. L. and Dittmer, D. S. (1964). "Biology Data Book" p. 264, Federation of American Societies for Experimental Biology, Washington D.C., U.S.A.

Andersen, N. F., Delorme, E. J., Woodruff, M. F. A. and Simpson, D. C. (1959). *Nature, Lond.* **184**, 1952.

Bollman, J. L. (1948). *J. Lab. clin. Med.* **33**, 1348.

Bollman, J. L., Cain, J. C. and Grindlay, J. H. (1948). *J. Lab. clin. Med.* **33**, 1349.

Boyd, W. C. (1966). "Fundamentals of Immunology", 4th edn, Interscience Publishers Inc., New York.

Falconi, G. and Rossi, G. L. (1964). *Endocrinology*, **74**, 301.

Farris, E. J. and Griffith, J. Q. (1966). "The Rat in Laboratory Investigation", 2nd edn, Lippincott, Philadelphia, Pennsylvania.

Freund, J. (1947). *A. Rev. Microbiol.* **1**, 291.

Freund, J. and Thompson, K. J. (1945). *Science, N.Y.* **101**, 468.

Goudie, R. B., Horne, C. H. W. and Wilkinson, P. C. (1966). *Lancet* i, 1224.

Harrington, F. G., Greaves, J. D. and Schmidt, C. L. A. (1936). *Proc. Soc. exp. Biol. Med.* **34**, 611.

Johnston, F. (1959). *Lab. Practice* **8**, 14.

Kahler, H. (1942). *J. natn. Cancer Inst.* **2**, 457.

Kassel, R. and Levitan, S. (1953). *Science, N.Y.* **118**, 563.

Lambert, R. (1965). "Surgery of the Digestive System in the Rat", C. C. Thomas, Springfield, Illinois.

Leonard, E. P. (1968). "Fundamentals of Small Animal Surgery", W. B. Saunders and Co., Philadelphia and London.

Liebelt, A. G. and Liebelt, R. A. (1967). *In*, "Methods in Cancer Research" (H. Busch, ed.) Vol. 1, p. 143, Academic Press, New York and London.

Mouzas, G. and Weiss, J. B. (1960). *J. clin. Path.* **13**, 264.

Newbould, B. B. (1965). *Immunology* **9**, 613.

Reinhardt, W. O. (1945). *Proc. Soc. exp. Biol. Med.* **58**, 123.

Ronai, P. M. (1966). *Transplantation* **4**, 208.

Sato, M. and Yoneda, S. (1966). *Acta endocr., Copenh.* **51**, 43.

Tata, J. R. (1967). *Biochem. J.* **104**, 1.

Temple, P. L. and Kon, S. K. (1937). *Biochem. J.* **31**, 2197.

Waynforth, H. B. and Parkin, R. (1969). *Lab. Anim.* **3**, 35.

The Use of "High Energy" Phosphate Compounds in "in Vitro" Studies on Protein Synthesis

P. S. TODD AND P. N. CAMPBELL

Department of Biochemistry, University of Leeds, Yorkshire, England

In 1945 Zamecnik and Keller showed that, in the isolated microsome preparation, the incorporation of ^{14}C-amino acids into protein was dependent not only on the presence of ATP but also on a compound containing high energy phosphate. Amongst the substances used were phosphoenolpyruvate (PEP), 3-phosphoglycerate (3-PGA) and phosphocreatine (PC). Since PEP proved to be as satisfactory as PC in their case and since there were some troubles concerning the purity of PC, PEP has been used very extensively since that time. It is, of course, assumed that the purpose of the PEP is to regenerate ATP either from ADP resulting from ATPase or from AMP resulting from amino acid activation. Thus it is usual to add pyruvate kinase to the system.

In 1964, Munro, Jackson and Korner showed that for protein synthesis polysomes did not require the addition of an ATP-generating system provided higher concentrations of ATP were used together with extra Mg^{2+} ions to prevent the added ATP reducing the effective Mg^{2+} ion concentration. Nevertheless most workers still use PEP in such systems.

We have re-investigated the requirement for an energy-generating system in preparations of microsomes and polysomes from adult rat liver. A major disadvantage in the use of PEP is its high cost which is about £15 per g compared with about 15 shillings per g for 3-PGA.

I. MATERIALS AND METHODS

Uniformly labelled [^{14}C]leucine with a specific radioactivity of 300 mc/m-mole was used. PEP was the silver-barium salt prepared by the method of Clark and Kirby (1963). 3-PGA was the barium salt and PC was the disodium salt. Creatine phosphokinase was obtained from Sigma and pyruvate kinase from Boehringer. GTP and ATP were the sodium salts.

Ag and Ba were removed from the preparation of PEP by HCl and K_2SO_4 and the solution of free acid was adjusted to pH 7·3 with 5N KOH. The Ba was removed from the 3-PGA preparation with K_2SO_4 and the pH adjusted to 7·3 with 0·25N KOH. The disodium salt of PC was dissolved in water and adjusted to pH 7·3 with 0·1N HCl.

Washed microsomes were prepared as described by Campbell, Serck-Hanssen and Lowe (1965). Detergent-treated polysomes were prepared as described by Webb, Blobel and Potter (1964). Microsomes or polysomes were suspended in a medium containing $MgCl_2$ (10mM), KCl (25mM), Tris buffer, pH 7·8 (35mM), and sucrose (0·15M). The cell sap from the microsome preparation was passed through a column of G-25 Sephadex as described by Munro et al. (1964).

Incubations were carried out at 37° in air with gentle shaking. Unless otherwise specified each incubation tube contained the particle suspension (0·4 ml), cell sap (0·2 ml), [^{14}C]leucine (0·125 μc), GTP (0·25 μmole), ATP (2 μmoles), and an energy-generating system in a total volume of 1 ml. The latter consisted of either 10 μmoles PEP plus 50 μg pyruvate kinase, or 20 μmoles PC plus 100 μg phospho-creatine kinase or 10 μmoles 3-PGA.

After incubation 5 ml of ice-cold water was added to stop the reaction. From each tube 1·5 ml was removed and 0·5 ml 4N NaOH added. The mixture was left 5 min at room temperature to hydrolyse the RNA. To each tube was then added 1 ml trichloroacetic acid (50%) to precipitate the protein. The protein was collected on Whatman GF/A glass fibre discs by filtration. The precipitate was washed with 3 × 5 ml portions of trichloroacetic acid (5%) followed by 3 × 5 ml portions of fat solvent (chloroform, ether, methanol, 2 : 1 : 1). The filter discs were placed in scintillation vials and dried in an oven at 70° for 10 min, allowed to cool and to each was added 10 ml of the scintillation fluid of Mans and Novelli (1961). The samples were counted in a Beckman LS 200 B liquid scintillation spectrometer.

RNA was estimated by the method of Mejbaum (1939).

II. RESULTS AND DISCUSSION

Table 1 shows a comparison of the three energy-generating systems when a microsome preparation was incubated for 40 min. From these results the comparative activity of the three compounds is PEP≫ 3-PGA > PC. In the original report of Zamecnik and Keller (1954) the order was $PC = PEP \gg 3\text{-}PGA$. In order to demonstrate the effect of the presence of added ATP a microsome preparation was incubated for 40 min with 10 μmoles of PEP in the presence and absence of 2 μmoles of ATP. In the absence of ATP there was no activity over the unincubated control.

For polysome preparations (Table 2) the incubation was extended to 70 min since they retain their activity for a longer time than microsome

TABLE 1

Effect of energy-generating systems on the activity of a microsome preparation

System	Radioactivity of protein
Unincubated	27
No generating system	193
PC	3588
3-PGA	3909
PEP	6757

Each tube containing 4 mg of RNA was incubated for 40 min. Radioactivity is expressed as counts per min per mg RNA. Mg^{2+} ion concentration is 6mM in the assay system.

preparations (Korner, 1961). For polysomes PC appears to be a better energy source than 3-PGA or PEP. As with microsomes, the polysome preparation showed a virtual complete dependence on the presence of ATP.

TABLE 2

Effect of energy-generating systems on the activity of polysome preparations

System	Radioactivity of protein
Unincubated	47
No generating system	3,070
PC	27,860
3-PGA	17,020
PEP	17,760

Each tube containing 2·1 mg of RNA was incubated for 70 min. Radioactivity is expressed in counts per min per mg RNA. Mg^{2+} concentration is 6mM in the assay system.

The superiority of PC in the polysome system could have been due to the chelating effect of PEP and 3-PGA for Mg^{2+} ions. As Staehelin

(1969) has pointed out, 10 μmoles of PEP bind 5 μmoles of Mg^{2+}. When the concentration of Mg^{2+} ions in the incubation medium was increased from 6mM to 10mM the radioactivity of the resulting protein was increased from 12,000 to 13,300 in the case of PEP, and reduced from 18,200 to 10,000 in the case of PC. While, therefore, there was a stimulation of the activity in the presence of PEP it still fell short of the activity with PC. No doubt the inhibitory effect of the added Mg^{2+} ion with PC was due to the poorer chelating properties of this substance so that the effective Mg^{2+} ion concentration was much greater than with PEP.

The effect of omitting the energy-generating system and increasing the amount of ATP was then determined. Table 3(a) and (b) shows the results for the microsome preparation. The effect of adding ATP in the absence of additional Mg^{2+} ion was inhibitory. In the presence of added Mg^{2+} ion there was little effect in the presence of PEP but in the

TABLE 3

Effect of increasing amounts of ATP in the presence and absence of added Mg^{2+} ions on the activity of microsome preparations

(a) Addition of ATP alone at 6mM Mg^{2+} ions

ATP concn.	Without PEP	With PEP
2mM	349	10125
5mM	615	4718
10mM	288	305

(b) Addition of ATP and Mg^{2+} ions

ATP concn.	Mg^{2+} ions	Without PEP	With PEP
2mM	6mM	197	5590
5mM	9mM	393	5960
10mM	14mM	1040	4990

Conditions as for Table 1. Mg^{2+} ion concentration refers to the final concentration in the assay system.

absence of PEP stimulation was marked. However, in no case did the activity in the presence of ATP alone reach the level of that in the presence of PEP.

Table 4 records the result of similar experiments with polysome preparations. Again, with increasing ATP at a constant Mg^{2+} ion concentration there was inhibition in the presence and absence of PEP. With addition of ATP and Mg^{2+} ions there was little effect in the presence of PEP but there was a most marked stimulatory effect in its absence. In fact the claim of Munro *et al.* (1964) that a polysome preparation does not require an energy-generating system provided more ATP and Mg^{2+} ions are added is amply confirmed, for the activity in the presence of 10mM ATP and 14mM Mg^{2+} ion was greater without PEP than with PEP.

The effect of various concentrations of K^+ ions on the activity of the system was determined. It was shown that both in the microsomal and polysome systems further addition of K^+ from 30 μmoles to 180 μmoles was either without effect or was very inhibitory especially with polysomes.

TABLE 4

Effect of increasing amounts of ATP in the presence and absence of added Mg^{2+} ions in the activity of polysome preparations

(a) Addition of ATP alone at 6mM Mg^{2+} ions

ATP concn.	Without PEP	With PEP
2mM	4060	16,140
5mM	8668	3,094
10mM	440	422

(b) Addition of ATP and Mg^{2+} ions

ATP concn.	Mg^{2+} ion concn.	Without PEP	With PEP
2mM	6mM	4,060	16,140
5mM	9mM	10,300	16,800
10mM	14mM	21,400	15,650

Conditions as for Table 2. Mg^{2+} ion concentration refers to the final concentration in the assay system.

The utilization of ATP for peptide bond formation leads to the production of AMP and the presence of an energy-generating system alone will not convert the AMP to ATP. For such regeneration of ATP one requires myokinase. We therefore determined whether the liver cell sap used contained this enzyme. We found that the activity was 0·74 μmole AMP transformed to ATP per min by 0·2 ml of cell sap. Thus there is at least potentially sufficient myokinase in the liver system. It might be well to check this point in other systems.

In conclusion, therefore, it seems that in microsome preparations it is necessary to include an energy-generating system and that PEP is the most active but that in polysome preparations an energy-generating system is not required provided more ATP and Mg^{2+} ions are added. In any case PEP does not seem to be the substance of choice in polysome preparations.

This work was supported by a grant from the Medical Research Council to one of us (P. N. C.).

REFERENCES

Campbell, P. N., Serck-Hanssen, G. and Lowe, E. (1965). *Biochem. J.* **97**, 422.
Clark, V. M. and Kirby, A. J. (1963). *Biochim. biophys. Acta* **78**, 732.
Korner, A. (1961). *Biochem. J.* **81**, 168.
Mans, R. J. and Novelli, G. D. (1961). *Archs Biochem. Biophys.* **94**, 48.
Mejbaum, W. (1939). *Hoppe Seyler's Z. physiol. Chem.* **258**, 117.
Munro, A. J., Jackson, R. J. and Korner, A. (1964). *Biochem. J.* **92**, 289.
Staehelin, M. (1969). *Biochim. biophys. Acta* **174**, 713.
Webb, T. E., Blobel, G. and Potter, V. R. (1964). *Cancer Res.* **24**, 1229.
Zamecnik, P. C. and Keller, E. B. (1954). *J. biol. Chem.* **209**, 337.

Author Index

Subject Index

A

Absorption rates
  of injected substances, 214
Acridines
  mutagenic effect, 2, 12
Actinomycin D, 92, 94
  blocking of transcription by, 92
Adjuvants, 245
  administration, 247
  preparing emulsions with antigens,
    246
Adrenalectomy
  of rat, 236, 237
Aggregate structure
  and amino acid incorporation acti-
    vity, 103, 104
Alum precipitate, 247
Amber mutants, 4, 19, 20
  classes of, 21
  suppressors, 23
Amino acid incorporating systems
  bacterial contamination of, 90
  directed by various polynucleotides,
    81
  effect of exogenous components, 88
  from cytoplasm, 77
  from plant organelles, 86
  in isolated chloroplasts, 89
  in mitochondria, 89
  incorporation of L-lysine, 78
  incorporation of phenylalanine, 79,
    80
  requirements, 77
Amino acids
  changes produced by mutation, 12,
    13, 14, 15
  codons for, 8, 10, 11, 82
  of proteins, 2
  replacement in mutants, 9, 10
    stepwise addition, 112
Aminoacyl-tRNA
  aminoacyl transfer, 85
  binding to ribosomes, 82, 83

Aminoacyl-tRNA (*cont.*)
  chromatography, 70, 95
  formation, 62, 66
  preparation, 68
Aminoacyl-tRNA synthetases, 60
  assay methods, 61, 62, 63
  properties, 63
  purification by chromatography, 63
  reaction mechanism, 61
  species specificity, 64, 65
Aminoacyl transfer enzymes, 69
  assay, 72
  chromatography, 73
  effect on polyphenylalanine forma-
    tion, 73
Anaesthesia
  induction, 229
  maintenance, 230
  tests for, 230
Anaesthetics
  barbiturate, 231
  volatile, 229
Animal experimentation
  integration with chemical analysis,
    210
Animals
  handling, 210
  operative techniques, 209
Antibody production, 245
Antigen, 245
  booster dose, 247
Antipolarity, 28
Arterial puncture
  of rabbit ear, 226

B

Bacteriophage
  coat proteins, 27
  control of development, 43
  head protein, 4, 6, 27
  T_4, conditional lethal mutants, 19
Barriers, 3, 22
Bentonite, 170